FATMA BELBEL
MEHDI BOUKHEROUFA
WAIL ZERIBI

Micromammal prey in the Edough forest massif

AF514428

FATMA BELBEL
MEHDI BOUKHEROUFA
WAIL ZERIBI

Micromammal prey in the Edough forest massif

Typology and monitoring through the study of the diet of the common genet and the African golden wolf

ScienciaScripts

Imprint

Any brand names and product names mentioned in this book are subject to trademark, brand or patent protection and are trademarks or registered trademarks of their respective holders. The use of brand names, product names, common names, trade names, product descriptions etc. even without a particular marking in this work is in no way to be construed to mean that such names may be regarded as unrestricted in respect of trademark and brand protection legislation and could thus be used by anyone.

Cover image: www.ingimage.com

This book is a translation from the original published under ISBN 978-620-6-72788-0.

Publisher:
Sciencia Scripts
is a trademark of
Dodo Books Indian Ocean Ltd. and OmniScriptum S.R.L publishing group

120 High Road, East Finchley, London, N2 9ED, United Kingdom
Str. Armeneasca 28/1, office 1, Chisinau MD-2012, Republic of Moldova, Europe
Managing Directors: Ieva Konstantinova, Victoria Ursu
info@omniscriptum.com

Printed at: see last page
ISBN: 978-620-8-50573-8

Copyright © FATMA BELBEL, MEHDI BOUKHEROUFA, WAIL ZERIBI
Copyright © 2024 Dodo Books Indian Ocean Ltd. and OmniScriptum S.R.L publishing group

Micromammalian prey in the Edough forest massif: Typology and monitoring through the study of the diet of the common genet and the African golden wolf

Authors: Fatma BELBEL ; Mehdi BOUKHEROUFA ; Wail ZERIBI

SUMMARY

At the heart of food webs, the diversity and dynamics of small mammals are excellent indicators of ecosystem health. Our research was based on a spatio-temporal analysis of the consumption of small mammals by two of their predators: the common genet and the African golden wolf. The resulting study was carried out in the Edough mountain range from December 2018 to February 2020, where we were able to collect a total of 390 droppings, including 210 belonging to the common genet and 180 to the African golden wolf. We carried out a taxonomic identification of the small mammals consumed by the two predators, then analyzed some parameters of the structure of the small mammal population, and its seasonal variation in natural environments. After processing the feces, we were able to identify 04 species of small mammals, with an overlap of trophic niches between the two predators, particularly marked in winter. These results confirm the status of the two sympatric predators as samplers of micro mammal biodiversity in natural environments.

Keywords : *Micromammals* - item - prey - *Genetta genetta* - *Canis anthus* - Massif Edough mountains.

CONTENTS

INTRODUCTION GENERALE

I. GENERAL INTRODUCTION

It is commonly accepted that the population, the set of individuals of the same species occupying the same ecosystem, is the fundamental elementary unit of ecological systems, since ecosystems are made up of interconnected populations **(Barbaut, 1994)**. However, they do not in themselves constitute functional units that can be studied in isolation. From a purely functional and evolutionary point of view, the smallest entity that a good ecologist can consider is what we call a "population-environment system" (Barbaut, Dajoz). **Figure 1** shows a simple food web, considered to be the typical structural core of ecological systems **(*In* Barbault, 1995)**: within the ecosystem to which it belongs, any population can interact with its environment in five main ways. On the one hand, it is involved in vertical interactions with its prey and predators or parasites, and horizontal interactions with other populations belonging to the same trophic level (competitive relationships). On the other hand, it is directly or indirectly subject to the effects of physical and chemical environmental factors, even if it may benefit from positive interactions (mutualism or symbiosis) with other species. Finally, intrinsic processes can also play a role in its dynamics. It is therefore clear that population-environment systems and food webs are representations of ecological assemblages that reflect the main interactions linking all the species present at a given time, in a given space.

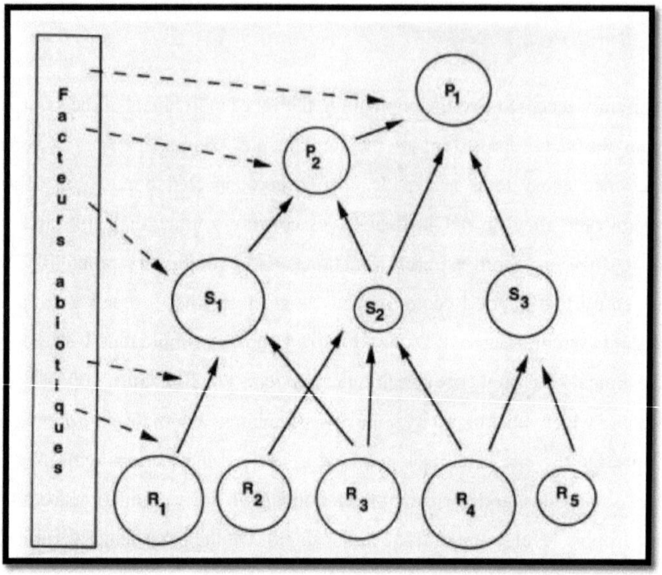

Figure1 : Representation of an ecological system centered on a food web of 3 species, sharing Ri resources and subject to two predatory species (*In* Barbault, 1995).

All current research has shown that predatory species, and Mammals in particular, are excellent bio-indicators of ecosystem health **(Wilson and Reeders, 1993)**. Mammals are predatory species at the top of the food pyramid. Consequently, their diversity, abundance or regular presence guarantees a high degree of prey diversity and, as a general rule, the maintenance of biodiversity within ecosystems **(Boukheroufa, 2018)**. However, following the extinction or near-extinction of large predators in North Africa, such as the Cheetah, Leopard and Atlas Lion, trophic competition levels between the remaining species (meso-predators) have been altered **(Aulagnier, 1992; Caro and Stoner, 2003; Naves *and al.,* 2003; Dalerum *and al.,* 2009)**. This trophic demotion could ultimately have tragic consequences for the ecosystem's equilibrium. Regular monitoring, through analysis of their diets , is therefore essential if management and conservation methods for mammal-prey systems are to evolve **(Klare *and al.,* 2010)**.

From a taxonomic point of view, Mammals is a highly diverse class, comprising not only terrestrial quadrupeds (the vast majority of Mammals), but also winged animals (Chiroptera or bats), aquatic animals (Pinnipeds [sea lions, seals...] Sirenians [manatees...] and even Cetaceans [sperm whales, orcas...]) ***(In* Boukheroufa, 2018)**. There are over 5,500 species currently living in the world, divided into 1,200 genera, 150 families and some thirty orders. Some specialists believe that 7,000 species are still unknown, some of them threatened with extinction **(Ceballos and Ehrlich, 2009)**. Species size varies considerably from group to group: the largest Mammals are marine and belong to the Whale family (the blue whale, Balaenoptera musculus). Among terrestrial Mammals, the Ungulates are the largest: the Elephant reaches 3 to 4 m at the withers; the Giraffe raises its head to over 5 m. The opposite extreme is found among Micromammals, including rodents, chiropterans and especially terrestrial insectivores; some shrews (such as *Crocidura etrusca*) are only 3 to 4 cm long and weigh just a few grams **(Kingdon, 2010; Aulagnier al., 2008, 2010)**. Mammals are therefore highly diverse in terms of species, behavior (diurnal or nocturnal, aquatic, terrestrial or aerial) and morphology (from a few centimeters to several meters, and from a few grams to several tons). What's more, some species are very rare or very abundant, while others are endemic to a specific region of the globe or inhabit every continent. This great diversity is often compounded by the very variability of the species' living environment, for although mammals display a wealth of complex behaviors, they also vary according to season, time of day and environmental conditions. All these features make the study of mammals highly complex, and explain the difficulties encountered in studying both individual and population biology. These difficulties are all the more accentuated when it comes to studying small mammals, the monitoring of which is a fundamental piece of data in population ecology, as they provide information on the resources offered by an environment, thanks to their strategic positioning at the heart of food webs **(Krebs et al, 2014)**. Micromammals are excellent environmental indicators, as they play a very important role in regulating animal populations (**Winemiller and Polis, 1996**). According to **Oppliger (2008)** and **Vincent (2000),** small mammals represent a group of species essential to the maintenance of terrestrial ecosystems, due to their predominance in the diet of a large number of predators and their high consumption of plant production. Small mammals are not a strictly defined group of species. The designation "small mammals" for these species, which belong to different taxonomic and phylogenetic groups, is based solely on their size. All terrestrial mammals smaller than or equal to the muskrat, i.e. up to 40 cm without tail **(Rigaux & Dupasquier, 2012; Wilson and Reeder, 2005)**, have already reported 3,845 species of small mammals, i.e. almost 70% of all mammals worldwide**.** 40% of these are rodents, making them the largest taxonomic order, followed by terrestrial insectivores**.** A recent survey of Algeria found 64 species of small mammals, including 30 species of rodents, 26 species of chiropterans and 8 species of terrestrial insectivores **(Ahmim, 2019).** Rodents include field mice, voles, mice and rats; terrestrial insectivores include shrews and hedgehogs; and chiropterans, commonly known as bats. Small mammals are found in almost all terrestrial environments (except the two poles), as they present a varied ecology in terms of both diet and lifestyle **(Mistrot, 2000)**. Most of them are active mainly in the twilight and at night, making them difficult to observe. This difficulty is compounded by the fact that most of them are also burrowers, digging galleries to build their nests **(Mistrot, 2000)**. These small mammals are prolific animals, although this character varies from one species to another, and their reproduction is at its peak from early spring to early autumn, i.e. when weather conditions are most favorable and resources are abundant **(Salder, 2012)**

Micro-mammalian diversity has been studied using two types of method **(Spitz, 1963).** In addition to direct trapping methods, as shown in **Figure 2**, many monitoring studies have been based on indirect methods, using the identification of prey found in predator diets. These micro-mammal prey items can be found in the droppings of birds of prey or in the droppings of carnivorous mammals **(Tanguy & Gourdain, 2011).** With carnivores in particular, several methods have been proposed to study the predation phenomenon and characterize the diversity of prey items, including small mammals **(Fedriani & Travaini; 2000).** Indirect or coprometric methods involve the collection of faeces, which has the advantage of not altering the structure of the predator population, unlike the collection of digestive tracts, for example, which requires sampling that is harmful to the fauna **(Day, 1966; Corbet, 1989; Mills, 1991; Damange, 1999).** It enables the proper identification of consumed

Micromammals, through the use of a simple protocol, good taxonomic characterization keys and appropriate laboratory equipment **(Quéré, 1993).**

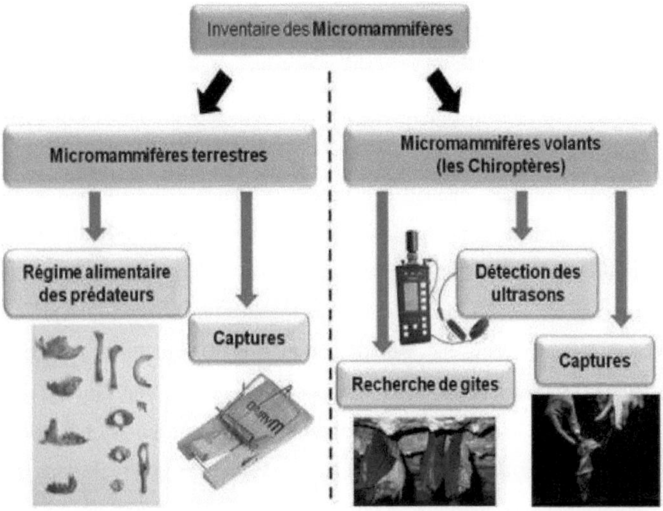

Figure2 : Summary of different methods for surveying terrestrial and flying Micromammals (Boukheroufa, 2018).

The monitoring of small mammal species through the analysis of their predators' diets is now described by many specialists as a non-destructive population conservation technique that can provide information on the diversity and abundance of these small organisms, their favorable or unfavorable habitats, as well as their direct relationships with their predators **(Davison et al., 2002; Tamnling et al., 2012; Torres et al., 2013; Drouilly et al., 2018**). Knowledge of the diet of these predators is therefore essential, as it enables us to understand the trophic preferences of prey items, their causes and the periods that are favourable for the consumption of small mammals.

In this study, we looked at micromammalian diversity in the diet of two mesopredators: the common genet *Genetta genetta* and the African golden wolf *Canis anthus*. The common genet (*Genetta genetta*) is one of the most extensively studied species across its entire range **(Lodé *et al.*, 1991; Palomares and Delibes, 1991;Gomes and Giraudoux, 1992; Carvalho and Gomes, 2004; Roberts et al, 2007; Palazon et al, 2008; Souret and Riols, 2018).** The genus *Genetta* is endemic to the African continent, with the exception of *Genetta genetta*, which is found in the Maghreb, the Arabian Peninsula and southwestern Europe, the Genetta having probably been introduced into the latter region during the historical period **(Lever, 1985; Morales, 1994; Amigues, 1999).** This meso-predator is especially renowned for its selectivity towards small mammals, making it a model of choice and an important tool for monitoring micro-mammalian biodiversity **(Boukheroufa *et al.,* 2020).** In Algeria, numerous authors have described the composition and dynamics of the common genet's diet, and have confirmed its selectivity towards micromammalian prey, even if the predator supplements its diet with other prey items **(Desmet et al., 1988; Delibes *et al.,* 1989; Hamdine *et al.,* 1993; Boukheroufa, 2005; Mansour *et al.,* 2005, 2014; Boukheroufa *et al.,* 2009; Bensidhoum, 2010; Ahmim, 2019, Boukheroufa *et al.,* 2020; Belbel *et al.,* 2022a, 2022b).** However, the selective, timid and shy nature of the common genet makes it fragile in degraded environments, which could compromise monitoring missions of micro-mammal populations. This led us to choose a second biological model with greater ecological plasticity, adapting perfectly to the anthropization of natural environments thanks to its eminently opportunistic trophic behavior. This is the African golden wolf *Canis anthus,* formerly known as the Golden Jackal (*Canis aureus*), which is now one of the top predatory species throughout its range (**Edinne, 2017**). Following eco-ethological studies, combined with genetic analyses undertaken between 2012 and 2015, on a large number of Old canids, it turned out that African Canids are more closely related to the gray wolf (*Canis lupus*) and the coyote (*Canis latrans*) than to the Eurasian golden jackal (*Canis aureus*). is why species *Canis aureus* Jackal) has seen its taxonomic classification evolve towards *Canis anthus* (African Golden Wolf) (Rueness, E. K et al., 2011; Gaubert, P et al., 2012; Stoyanov.S 2020; Krofel et al., 2021).

In Algeriathe African golden wolf remains largely unstudied, and the few existing studies concern, in particular, the diet of this predator, and have mainly been carried out the Djurdjura National Park in Kabylie **(amroun *et al.,* 2006, 2014)** and in the Edough mountain massif **(Boukheroufa *et al.*, 2020; Belbel et al., 2022 a, 2022b)**. These pioneering works the trophic opportunism of the species, which possesses a broad food spectrum anddepending on the trophic conditions of the environments in which it evolves, will adopt a more or less partial overlap in trophic niche it shares other predators such as the common genet *Genetta genetta.*

In the light of all these arguments, our problematic was built around the analysis of micro-mammalian biodiversity through the diet of the common genet and the African golden wolf, in different localities of the Edough mountain massif. This peninsula, located in the extreme north east of Algeria, is considered a regional biodiversity hotspot named "Kabylie - Numidie-Kroumirie" **(Véla & Benhouhou 2007)**, containing many important areas for plants **(Radford & al. 2011)**, with a wealth of endemic and sub-endemic taxa with fragmented ranges, rare in Algeria, often very localized **(Yahi & al. 2012; Hamel & al. 2013, 2022a, 2022b)**. The Edough peninsula is therefore an exceptional site in which trophic chain analysis could provide fundamental data on the state of equilibrium of the environment, but also on the concrete actions to be taken for the conservation of this hotspot with local, national and global influence.

The main objective of this study is to analyze the diversity and dynamics of micro-mammalian prey located at the heart of sympatric predators' food webs, which must adopt strategies to minimize the degree of trophic niche overlap in natural environments. To achieve this objective, we asked ourselves the following questions: ***Which small mammals do the two predators eat?*** To answer this question, we analyzed the diets of the two predators, and identified the small mammals, using dichotomous keys based on the observation of bones and cross-sections of hairs found in feces.

Of the two predators, which is the better sampler of micro mammal biodiversity???

To answer this question, we carried out a comparative analysis of the proportion of small mammals in the overall diet of the two predators.

MATERIALS AND METHODS

II. MATERIALS AND METHODS

Our study was carried out in the Edough mountain range, one of Algeria's most diverse areas, since it is considered to be a hotspot of plant biodiversity in the paleo-arctic ecozone, with a large number of rare species of great biogeographical interest, and a high diversity of fauna **(Comes, 2004).** We have therefore drawn on a large body of research to briefly describe the main features of the study area.

2.1 D escription of the study area

2.1.1 Geographical location

The Edough is a small coastal massif located in northeastern Algeria, west of the town of Annaba. It stretches between Cap de Garde to the east, which closes the Bay of Annaba, and Cap de Fer to the west of the massif, which delimits the Gulf of Skikda. The Edough massif is bordered to the south-east by the Guerbès Senhadja wet plain, a major Ramsar site in 2001 (**Toubal *et al*, 2014)**, and the Annaba plains, to the south by Lake Fetzara, to the west by the Kherraza plain, to the north by the Mediterranean Sea and to the north-west by the Chetaibi mountain range. The ridgeline stretches for 26 km, reaching 1,008 m at Kef Sebaâ, the highest point in the region, before dropping to 867 m towards the village of Seraidi and dropping to the tip of the Cap de garde peninsula north of Annaba **(Hani et al, 1997) (Figure3).**

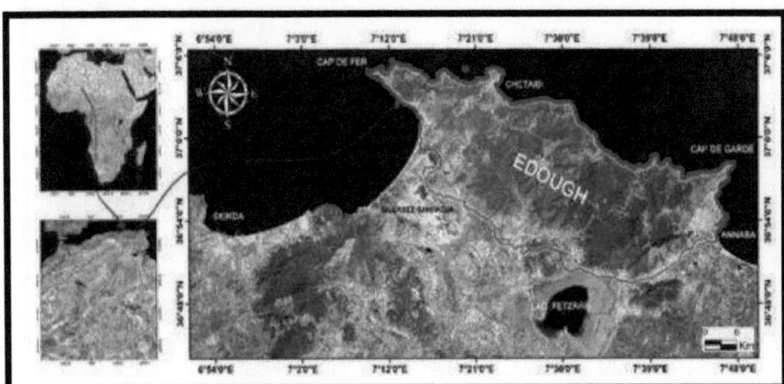

Figure3 : Location of study area (Present study).

2.1.2 Physical characteristics

-Geology

The Edough peninsula is a group of hills with altitudes ranging from 0 m to 1008 m, characterized by steep slopes. The Edough forms a crystalline dome. The heart of the dome is made up mainly of gneisses in complex association with ultrabasic rocks **(Bossière et al, 1978; Caby et al, 2001; Hadj Zobir et al, 2007),** topped by garnet, kyanite and staurotide micaschists associated with marbles topped by staurotides, andalusites, and quartzite beds with lenses of gneissic leucograite. The alternation of micaschists and quartzite, also known as the "alternation series", has been dated to the Lower Paleozoic by acritarchs **(Ilavsky and Snopkova, 1987; according to Hani et al., 1997).** The central part of the massif consists mainly of gneisses and migmatites intercalated with marbles and amphibolites. **(Figure 4).** Magnetite mineralization of meta-sedimentary origin is the new era in the formation of new deposits **(Henni and Aissa, 2007**).

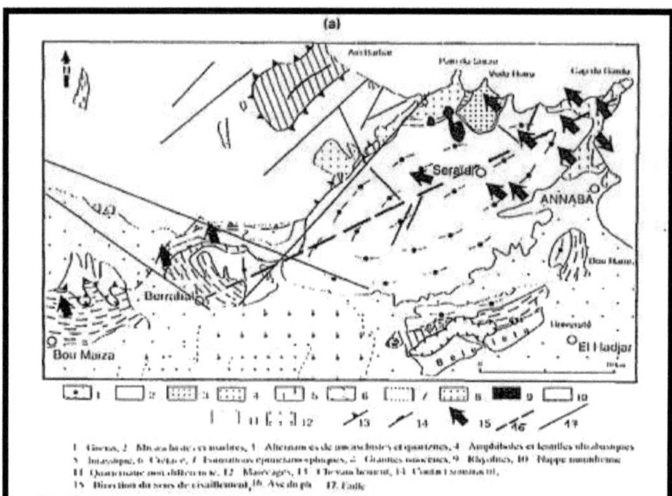

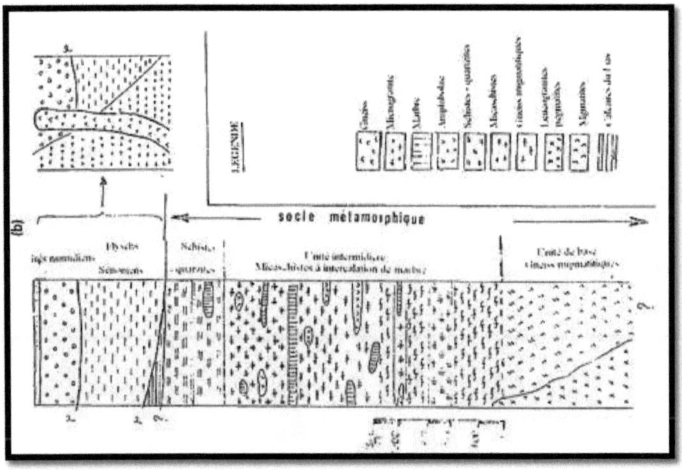

Figure4 : (a) and (b) represent the geological map of the major structural features and lithological column of the Edough massif *in* (Hani et al, 1995).

-Pedology

The crystallophyllous Edough massif corresponds to eruptive, metamorphic and sedimentary formations **(Toubal and Boumaza, 1989).** These acidic soils, derived from siliceous sea rock, are ideal for various acidophilous plant groups.

The soil is the result of several natural and anthropogenic factors (pedogenesis, land use), its potential importance stemming from its contact with water (water networks and rainfall) and air (aeration and weathering), allowing different plant groups to develop according to their physicochemical characteristics. Some of these, on the south-eastern side of the Edough in the forests, develop on brown or leached soils, sometimes with a podzolic tendency **(Toubal and Boumaza 1989; Toubal, 2013)**. According to **(Oularbi and Zeghiche, 2009).** Sensitivity to erosion is high, as pedogenesis rarely occurs, although morphogenesis is very active and results from the combination of steep slopes and violent downpours. The soils are poorly structured, but with good vegetation cover, favoring morphogenesis.

-Hydrology

The Edough peninsula comprises four watersheds more or less bounded by the main ridge lines. The hydrographic network is very dense and drainage is dendritic. The steep slopes and the nature of the rock do not allow for diffuse runoff and favor rapid drainage. Rainfall is abundant, often exceeding one meter **(Hadj Zobir, 2012).** These characteristics influence the region's susceptibility to erosion. **(Oularbi and Zeghiche, 2009) (Figure 5).**

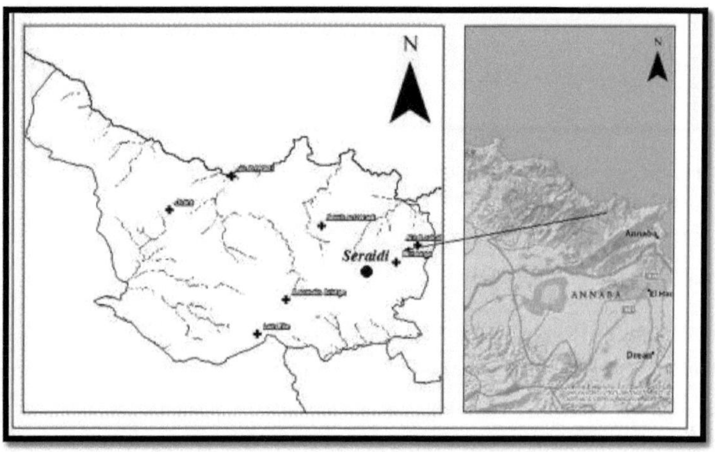

Figure5 : Dendritic drainage network of the Edough peninsula (Oularbi and Zeghiche, 2009).

-Climate

The region studied is characterized by a Mediterranean climate, with two distinct seasons, one cool, humid and rainy, the other hot and dry **(DGF, 2002).** Due to its high-altitude location, the region receives abundant precipitation. Snowfall is also frequent, accompanying rain, mist and hail, and covering peaks above 700 m in altitude. Snow cover can reach 20 cm **(Hani et al., 1997)**.

-Precipitation and temperatures

Rainfall varies according to an altitudinal gradient, from sea level to 1008 m, the highest point in the region. Precipitation is very abundant, often exceeding 1 metre annually.

Temperature variations follow several criteria, depending on season, latitude, altitude and atmospheric conditions. In the Edough peninsula, the highest average monthly temperatures are observed during the period from May to October, while the dry season currently extends from June to September, and the lowest temperatures are observed from November to early March.

Ground cover

Thanks to their acidity and humidity, the soils favor the formation of a mosaic of cork oak (*Quercus suber*) **(photo 1),** and zen oak (*Quercus faginea*), a very important subéraie. The Edough is also characterized by chestnut (*Castanea sativa*) and maritime pine (Pinus *pinaster*), although it is more tolerant of limestone.

photo1 : Vegetation cover of the Edough massif (© Belbel).

These formations, with their affinity for calciferous soils, also allow their flora to establish themselves, with Ericaceae (such as Arbutus *unedo*, tree heather (*Erica arborea*), Limaceas such as (*Lavendula stochas*), Cistaceae such as (cistus monspeliensis, calicotome villosa) and Ascteraceae such as (galactite tomentosa), Cistaceae such as (*cistus monspeliensis, calicotome villosa),* Ascteraceae such as (*galactite tomentosa*), Caprifoliaceae such as (*Knauta arvansis*), Leguminosae such as (*cystus villosus*), Rosaceae such as (*Rubus fruticosus.l*) ,Oxalidae such as (*oxalis pes-caprae*), Linaceaceae such as (*Linum usitatissimum.L*) **(Photo 2).**

photo2 : a few plant species from some of the sites sampled (© Belbel)

Vegetation is distributed according to local ecological conditions: altitude, topography, substrate, bioclimate and stages. The altitude of the Edough massif is characterized by several vegetation strata arranged in 3 floristic altitudinal stages:

- The Mediterranean thermo stage from 0 to 500 m example
- The Mediterranean meso stage from 500 to 800m,
- The supra-Mediterranean stage above 800m.

The delimitation between strata is not a firm law, and it can happen that several formations find themselves in favorable conditions allowing cohabitation.

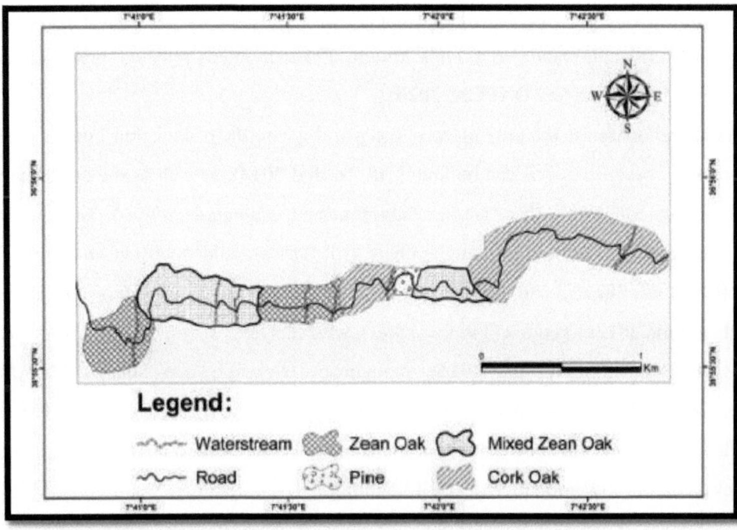

Figure6 : different vegetation successions at the Ain boukal site (*in* Laref et al., 2022)

-Rich fauna

The Edough mountain range is home to a particularly interesting diversity of animals. The presence of the Atlas bear (***Ursus crowtheri***) was reported in the Edough by the Spaniards, who occupied the citadel of Annaba in 1535 **(Wilson and Reeder, 2005)**. In 1847, Guyon admitted the existence of the Bear based on the testimony of the painter Vernet, who had seen a bear skin in Bône (Annaba) from an animal killed in the nearby mountains (**Ahmim, 2019)**. Algeria's last Atlas or Barbary lion (***Panthera leo leo***) was killed in the Edough massif in 1890 **(DGRF, 2006)**, where it is thought to have disappeared during the destruction of forests north of Sétif in 1958 **(Simon et al, 2013)**(**Djazairess ,2023**). A few specimens remain in captivity in some zoos in Europe and Morocco. Panthers also lived in forests before becoming extinct (**Ahmim, 2019**). The black panther has also lived in Algeria and North Africa, and was spotted in the Hoggar and northern forests in 2007 **(Henschel et al, 2008).** A few accounts have revealed its presence in the Kabyles forests (Bejaia) in 2021, while another researcher was surprised by its presence in El Kala in 2022. The cheetah still exists, but the species is critically endangered. Only around 200 individuals remain, spread across Algeria, Mali and Niger **(ONPCA, 2020).**

The Edough massif is the only place in the world where the endangered Poiret's grey newt (*Pleurodeles poireti*) can be found **(in Toubal 2014),** as well as the only genus of toothless collembola the Algerian **Salamander** (*Salamandra algira*)**.** The current common fauna found in the Edough consists of: African golden wolf (*Canis anthus*) **(Belbel et al., 2022b**), common genet (*Genetta genetta*) **(Boukheorufa et al., 2020; Belbel et al., 2022a)**, red fox (*Vulpes vulpes*), wild cat (*Felis silvestris*), wild boar (*Sus scrofa*) **(Benotmane et al., 2024)**, mongoose (*Herpestes ichneumon*), Algerian hedgehog *Atelerix algiru* ***(Sakraoui et al., 2014)*****,** porcupine (*Hystrix cristata)*. Forest birds in the Edough include blackbird, great spotted woodpecker, tits (blue, great and black), oak jay and wood pigeon **(DGF[i] , 2006).**

2.2. Presentation of biological models

The diversity of small mammal populations was studied by analyzing the diet of two distinct predators. These were the common genet *Genetta genetta,* and the African golden wolf *Canis anthus*.

2.2.1. The common genet

2.2.1.1. Systematics

Called 'Karnit' or 'Zirda' by local residents in the town of Séraidi, its scientific name is '*Genetta genetta'*, and its common name is 'la genette commune' **(photo 2).** The Viverridae family is of particular interest, as it enables us to learn more about the origins of carnivores. The genet *Genetta genetta* **(Linnaeus, 1758)** occupies the following position in the vertebrate phylum:

Order: Carnivora

Suborder: Feliformia

Family: viverridae

Subfamily: viverrinae

Genus: *Genetta*

Species: *Genetta genetta* **(Linné ,1758)**

Common genet Genetta genetta (Linnaeus, 1758)

photo3 : Common genet (El Kala - Algeria in April 2005) (© Raouf Boulahbal).

According to **(Schlawe,1980,1981),** in **(Livet et Roeder ,1987)** the genus ***Genetta*** comprises 10 species. All are African, only ***Genetta genetta*** is found in Europe. This taxon comprises five subspecies:

- *Genetta genetta* **(LINNE ,1758)** : Iberian Peninsula
- *Genetta genetta afra* **(Cuvier ,1825)** : North Africa (Morocco, Algeria, Tunisia, Libya, Egypt, Mauritania)
- *Genetta genetta balearica* **(Thomas ,1902**): Mallorca and cabrera
- *Genetta genetta isabelae* **(Delibes, 1979)**: Island of Ibiza
- *Genetta genetta rhodonica* **(MATSCHIE, 1902)** : Pyrenees, France.

2.2.1.2. Geographical distribution

According to **Livret and Roeder (1987),** the viverridae family originated in Ethiopia, which explains its African and southern European distribution. The common genet occupies a variety of habitats in its native range, from the savannahs of sub-Saharan Africa to the forests of the Maghreb and the arid coastal regions of the southern Arabian Peninsula **(Gaubert, 2009)**

- In Europe

The common genet has established itself in forested areas of the Iberian Peninsula (Portugal and Spain, including the Balearic Islands) and France (except Corsica) **(Gaubert, 2007).** A few individuals, probably fugitives, have been observed in Germany, Belgium and Switzerland. Most French genet populations are confined to the south of the Loire and west of the Rhône **(Croquet, 2005). (Figure 7).**

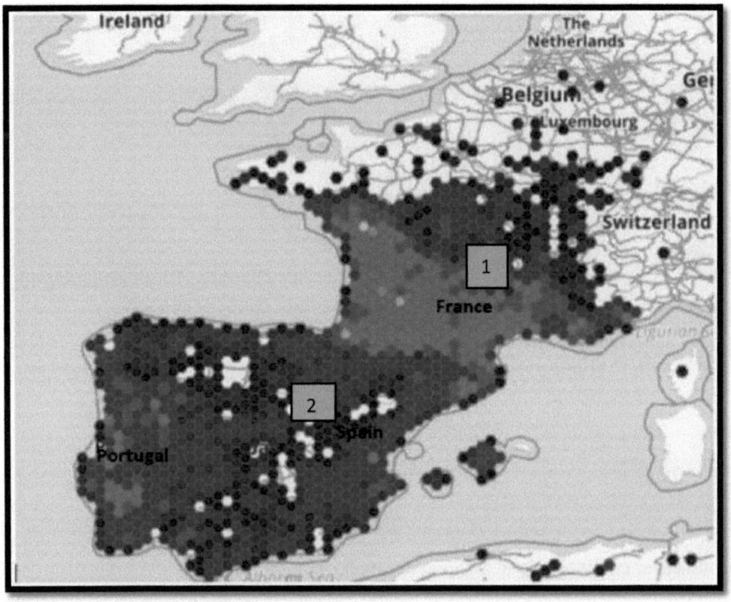

Figure7 : geographical distribution of the common genet (Cuvier 1825) in Europe from 1824 to 2022 (Belbel, present work) https://www.gbif.org ; ***(1) Genetta genetta genetta, (2)Genetta genetta rhodanica.***

- In Africa

The Genet is a species of African origin. It is widespread throughout the continent, with the exception of the desert and tropical forests. Most widespread in North Africa and sub-Saharan regions **(Mallil, 2012). (Figure 8)**

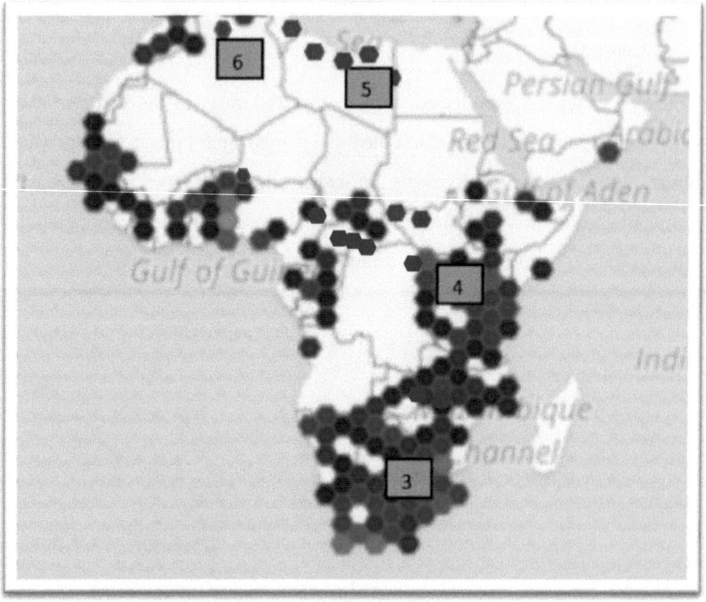

Figure8 : geographical distribution of the common genet (Cuvier 1825) in Africa from 1824 to 2022(BELBEL.F) https://www.gbif.org *(3) Genetta genetta pulchra (4) Genetta genetta felina (5)Genetta genetta balearica (6) Genetta genetta afra.*

- In Algeria

Its absence has been reported in steppe areas as well as in the Saharan atlas by **(Hamdine *et al,* 1993)**, but it is found in forest areas **(Hamdine, 1991 ; Amroun, 2006 ; Bensidhoum, 2010 ; Mallil, 2012 , Boukheroufa et al, 2020, Belbel et al, 2022).** Widespread in the north, it has been recorded in all national parks from Tlemcen to El Kala, including the Belezma national park . Early records mention it in the Saharan Atlas and (**Tristan 1960 in Kowalski and RzebikKowalski, 1991**) observed a genet between Laghouat and Djelfa in the list of animal species captured for parasitological studies in 1908 **(In Ben sidhoum, 2010 and in Mallil, 2012)**. Many authors have contributed to the historical distribution of genet species (**Corbet, 1966; Harrison, 1968; Pringle, 1977; Schlawe, 1981; Gasperetti, et al., 1985; Croquet, 2005; Gaubert, 2007; Mallil, 2012)**

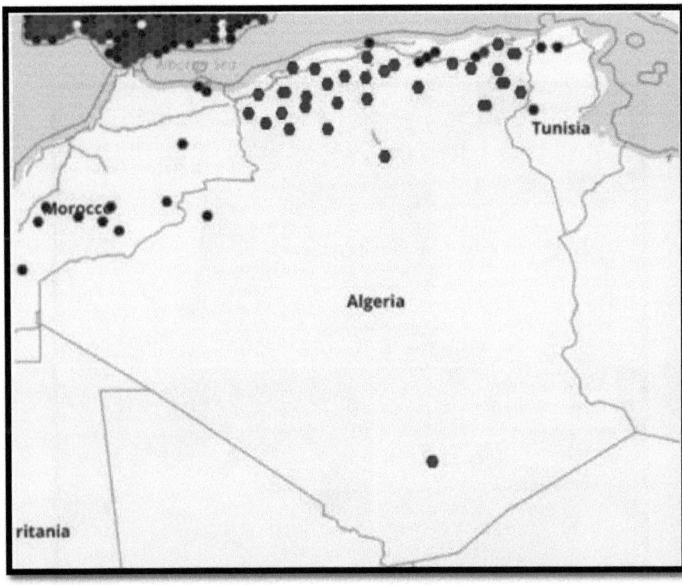

Figure9 : geographical distribution of the common genet (Cuvier 1825) in Algeria from 1824 to 2022 (Belbel,2023) https://www.gbif.org/species.

2.2.1.3. Habitat

Generally speaking, no study has proposed a typical Genet habitat. However, all the studies carried out on both shores of the Mediterranean basin show that this small carnivore displays great flexibility in its choice of habitat, depending on the ecological characteristics of the environment; For example, the concept of "rocky areas" recurs regularly in the literature, the proximity of waterholes, and that of "tall, dense plant formations", with a high availability of potential prey, safe shelter and low anthropization of the environment **(Lozé, 1984; Livet and Roeder, 1987; Maizeret**

et al, 1990; Hamdine, 1991; Ariagno, 1985; Virgos and Casanova, 1997; Amroun, 2006). However, some works have mentioned that the Genet can occupy sparsely vegetated areas, with dry zones **(Delibes, 1974 in livet et Roeder, 1987; Amroun, 2005)** and even woody arid steppes and Saharan steppes are Genet habitats in Morocco and at the El Mergueb reserve in Algeria **(Cuzin, 2002, Bensidhoum, 2010) (photo 6 and 7).**

photo4 : Different habitats selected by the common genet at Edough

(© Belbel, present work)

2.2.1.4. *Morphology of the common genet Genetta genetta*

- ***Morphological characteristics***

The genet is a small carnivore whose morphology makes it easy to identify. The coat and appearance are reminiscent of the cat (**Aulagnier and Thevenot, 1986**), but the body is more slender, the muzzle more pointed, the legs shorter and the neck and tail longer. The high-contrast coat is tawny-grey mottled with black-brown on the flanks in four to five longitudinal lines, with a black stripe on the upper back. The tail is almost as long as the body, ringed in light and dark. The genet has large ears that stand out well from the coat, and its elongated muzzle ends in a dark-brown tuft. Its total length is around 90 cm, with a tail of around 40 cm. Adult weight is between 1.5 and 2 kg (Léger, Ruette, 2010).

- ***Voice :***

Contact cry "Uff, Uff, Uff", excited growl, threatening hiss and spitting sound, contented purring during copulation, or at a comfortable rest.

- ***Body measurements :***

T+C= 40-55, Q = 40-51, HGT = 15-20, P = 3 kg. (Ahmim, 2019) according to our nocturnal observations we encountered genets of larger size at El Kala than at Edough (Seraidi).

- ***The genet's teeth***

Examination of the dentition shows that this family is not very advanced in the order of carnivores. Indeed, there are 40 teeth distributed according to the formula: I: 3/3; C: 1/1; PM: 4/4; M: 2/2 (Mallil, 2012).

2.2.1.5. *Traces and signs of presence*

The genet has five toes on each paw and claws that are semi-retractable, not visible and do not mark on the prints. The dimensions of its prints are according to **Boukheorufa, (2018) and Ahmim, (2019)** of: 25 - 42mm long and

30 - 63mm wide for the Pate anterior (PA), while 30 mm - 65mm long and 23 - 30mm wide for the Pate posterior (PP). The four upper digits are evenly fanned out above the plantar pad; the smaller fifth digit is clearly eccentric. Frequently, the fifth finger does not mark, so the genet's track resembles that of a cat **(Chazel, 2008; Boukheroufa, 2018; Livet and Roeder, 1987; Croquet, 2005).** Confusion between wolf and fox droppings, for example, is not common, since fox droppings are smaller, rarely contain bone fragments, are composed mainly of small mammal remains and fruit residues, and often contain insect fragments **(Bang, P., & Dahlström, P. 2001; Mech, L.D., & Boitani, L. ;2003; Elbroch, M. 2003).**

Photo5 : genet droppings, crottier (© Boukheroufa)

photo6 : Common genet *Genetta genetta* footprint (© Belbel).

2.2.1.6. Sexual dimorphism

The perineal glands, located between the anus and the vulva or penis, are extreme and crossed in their middle by the urogenital orifice; these glands are of the sebaceous type **(Livet and Roeder ,1987).** Anal glands are internal and abdominal, located on either side of the rectum. They open to the outside via a short canal at the level of the anal bulge. They consist mainly of sweat and apocrine glands and a few sebaceous islets, and are likely to coat the animal's faeces with their secretions **(Souloumiac and Canivenc, 1976; Livet and Roeder, 1987).** Plantar glands are located on the soles of the tarsi and metatarsi.

2.2.1.7. Reproduction

According to **(Livet and Roeder, 1987),** the male genet has no typical seasonal sexual cycle, and the testes maintain a complete lineage throughout the year. The genet rut occurs in January-February, with a secondary rut in May-June. These periods are not fixed, however, and births can occur all year round. Gestation is 55 to 77 days with copulation lasting 3 to 5 minutes, one to two litters per year of up to young and late sexual maturity (2 years) **(Ahmim, 2019)** . Although little is known about population dynamics, including survival rates by age class and sex, the genet is one of the species with a slow reproductive strategy like the weasel and marten **(Croquet, 2005; Léger and Ruette ,2010).**

2.2.1.8. Social organization and behavior

Genets are mainly solitary animals, although the ranges of males and females may overlap. Individuals of the same sex have exclusive territories. They are nocturnal and rarely appear during the day, although they are sometimes reported to forage at dusk. They are stealthy hunters, rather like cats, and kill with a quick bite to the neck. Their sharp claws enable them both to corner their prey and to be excellent climbers. Once their young have left the crottier, they repeat the same cycle, moving to a new territory or taking over the old one.

On forest sites regularly frequented by hikers, it takes advantage of the night after campfires have been extinguished to explore the remains of food left behind. Its shy behavior makes it sensitive to human presence and classifies it as an 'urban avoider' species (Blair, 2011).

2.2.1.9. Diet e

The Genet is defined as a generalist carnivore with a broad trophic spectrum, due to its marked opportunistic character **(Belbel et al., 2022).** Diet has been one of the best-studied aspects in various localities: (**Souret and Riols, 2018)** in France; **Virgos et al. (1996), (Torre et al. 2013; Palazon et al. 2008)** in Spain, **(Vingada et al, 1993), (Carvalho and Gomes, 2004; Rosalino and Santos.Reis, 2002)** in Portugal; (**Delibes et al. 1989)** in North Africa (Algeria and Morocco); **Roberts et al. (2007)** in South Africa; (**Hamdine et al, 1993**; **Amroun.., 2006; Bensidhoum, 2010)** in Kabylia (Algeria). **(Boukheroufa et al, 2009)** in EL Kala (El Tarf, Algeria). **(Belbelet al,2022)** in Edough, Annaba, Algeria. The results of diet analysis vary from region to region, as diet composition is influenced by several variables such as food availability, habitat type and season **(Vingada et al., 1993; Virgos et al., 1999).** Consistent with most results, small mammals represent an important part of the Genet's diet, and are mainly represented by rodents during wet seasons.

2.2.1.10. Conservation status

The common genet is classified as Least Concern (LC) on the IUCN Red List of Threatened Species. Although there are no specific conservation strategies in place for the genet, it is found in numerous protected areas throughout its distribution. The mortality rate has recently risen as a result of road accidents and illegal hunting in northern regions, and especially among farmers who own hen houses (personal observations).

2.2.2. *African golden wolf*

Vernacularly known as 'Dib', not only in the Seraidi or Annaba region but throughout northern Algeria, except in Kabylia, where its Berber name is 'Ouchane', hence the origin of a family name specific to the region. Its scientific name is *Canis anthus* and its common name is the African Golden Wolf.

2.2.2.1 *Systematics*

Order carnivora
Suborder caniformia
Family canidae
Caninae subfamily
Genus canis
Species Canis *anthus*

Photo7 : photos of the African golden wolf (©BELBEL.F)

2.2.2.2 Geographical distribution

- **In Edough**

We found these burrows next to a private farm located in the Bouzizi landfill. These burrows, designed to store food, are of varying depths, and are used as windbreaks and for sunbathing, according to testimonies from the landfill's guards and a former forest hunter. In addition to loopholes in farm buildings. These burrows are not to be confused with those dug by wild boar. The latter are shallow and represent the result of foraging, often for glasses of land, crops or favorite wild plants **(Photo 10).** We found its skull and coat at the end of the Bouzizi landfill. Given its urban nature, every evening a flock consisting of a couple and three young are regularly spotted near garbage cans in the village of Seraidi, and individuals are found dead on the roads leading from Seraidi to the town of Annaba.

- **In Algeria**

In Algeria, the African golden wolf is widespread throughout the country, from the coast to the southern limits. It can be found in the mountains of the central Sahara (Hoggar, Tassili N'ajjer). It can be found up to 2200 m altitude in Kabylie **(khidas 1998).** It can be found both on the plains and in the mountains **(Judas, 1998 ; Amroun, 2006**). It frequents all the biotopes found in the region: forests. Maquis, open areas, cultivated land and urban environments, where it requires a minimum of vegetation cover to hide .

2.2.2.3 Worldwide distribution

- ***Habitat***

The golden wolf occupies a wide variety habitats, ranging from semi-desert to savannah, woodland and mountain **(poché et al, 1987; fuller et al, 1989; kowalski and Rzebik-kowalska m, 1991; khidas et al, 1993; amroun et al, 2006).**

The African golden wolf is adapted to all types of habitat, which is why it is widespread across all Algerian landscapes, from the Sahara El Hoggar to Illizi and the far north of Algeria (the entire Mediterranean coast), in natural, semi-urban or even urban landscapes. It can easily adapt to human activity, which classifies it as an urban-adapted species.

2.2.2.4 Morphology of the African golden wolf Canis anthus

- Body measurements: Body measurements: 65 - 75 in length and 23 - 27 in width for the Pate antérieure (PA), as well as 30 - 41 in length and 50 - 55 in width for the pâte posterérieure (PP) according to **(Chazel , 2008; Boukheroufa, 2018).**
- Dental formula: the dental formula for the Canidae family is I3/3 + C1/1 +PM4/4 +M1-4/2-4. For the wolf, it is I3/3 + C1/1 + PM4/4 + M2/3. All African genera have 42 teeth. The canines are developed and exceed the other teeth **(Ahmim.M, 2019) .**
- Voice: The African golden wolf is a very vocal species, using a variety of calls such as barking as a warning sign. The most characteristic call is the howl of lament, often heard in chorus at dawn and dusk. Howling in adults (also in juveniles, but shorter and less loud): the contact call is a plaintive, ascending "Aouuu" howl lasting 5 to 8 seconds, often repeated. A shrill cry of greeting and excitement when hunting or fleeing from several members in danger. Wolves are adapted to twilight life, where visual signals are not possible. They emit sounds when searching for a mate or when hungry or cold, and other high-pitched cries when they throw themselves on their backs as a sign of submission, barks as alarm calls, purrs similar to those of the domestic cat when they are comfortable.

2.2.2.5 Traces and characteristic indices

The most obvious signs of presence after visual observation of the animal itself **(Photo 9)** or its corpse are its paw prints on the ground **(Figure 10**) or its faeces **(Photo 11; Photo 12).** Confusion between its footprint and those of a domestic dog in the field can be eliminated on the basis of the main characteristics mentioned **(Table 1).**

Table1 : Comparison between the presence of domestic dogs and African golden wolves.

Golden wolf	Dog

Fine, sharp claws	Large, rounded claws
Oval-shaped impression	Round impression
Lines can be drawn that do not intersect the bearings (Figure 11).	Not feasible
Forward-facing claws	Star-shaped claws
A gallop of imperfect straightness	A collected gallop

photo8 : Dress of a completely decomposed (decapitated) African golden wolf, Bouzizi site, February 22, 2020 **(© Belbel)**

photo9 : On the left a wolf's droppings in a tuft of the plant *Ampelodesmos moritanica* (El diss) pendqnt lq sqison sèche, Piste Oued El Gueb Avril 2018,; On the right fresh droppings covered with water droplets of African golden wolf site Ain Boukal fevrier 2018 **(© Belbel)**

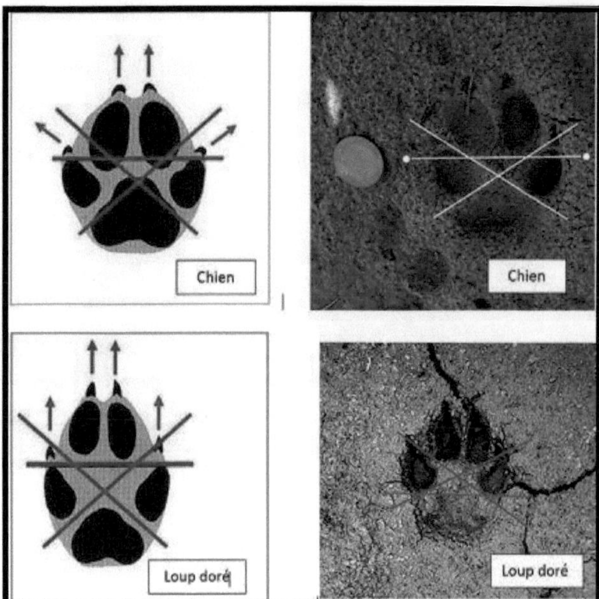

Figure10 : Visual comparison between the footprint of an African golden wolf (bottom) and that of a domestic dog (top) **(Belbel, present work).**

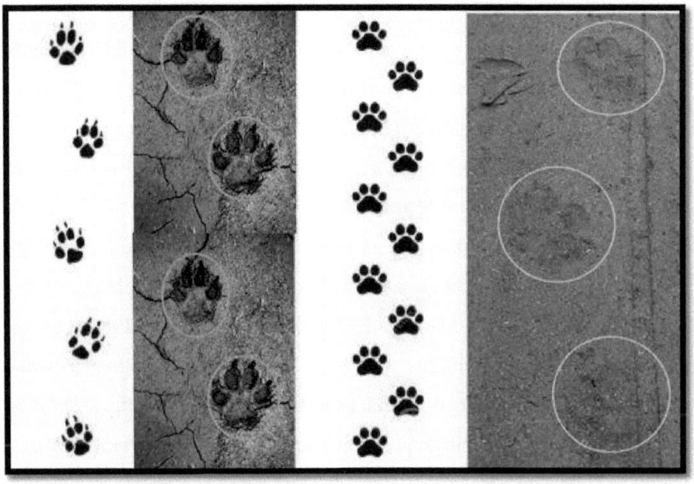

Figure11 : comparison between dog and wolf gallops(**© Belbel)**

2.2.2.6. Sexual dimorphism

For golden wolves, as for all carnivores, males tend to be generally larger than females, with more developed jaw musculature and larger upper canines **(Ewer, 1973).** For **khidas (1986),** the distinction can be made on the basis of morphological traits. Females have a finer, more pointed snout, giving the head a broader appearance than in males, and the belly of females appears heavier. The latter is the one most widely adopted by authors working on carnivores.

2.2.2.7. Reproduction and longevity

The female gives birth to 2 to 12 young in the spring, after a 2-month gestation period, from June to November, in a den (abandoned burrow) or a converted shelter (in a cave, bushes, under rocks, etc.). The group is made up of parents and young, who leave after 1 to 2 years. Gestation lasts an average of 63 days, with annual litters possible but not regular. The pups, numbering from 1 to 10, are born in burrows, holes or under a rock or bush, open their eyes at 8 days, leave the burrow at 14 days, and are suckled for 8 to 10 weeks. The longevity of the wolf is 10 to 12 years, and up to 14 in captivity. Between birth and weaning, pups depend entirely on their mother's milk. Both parents also help socialize the pups and, once weaned, regurgitate food for them. (**Moehlman, 1987; Moehlman and Hayssen, 2018**). The Golden Wolf is monogamous, with the breeding pair occupying a territory that is regularly marked and defended against intruders **(alden et al, 1996 and Macdonald, 2006).**

According to **Macdonald (2006),** litters generally occur during periods of high food availability. Sometimes the female may have two litters in one, but this is not a regular occurrence, as reported by **Haltenorth and Diller (1980).** Sexual maturity is reached at 10 months, with pairs forming from November and copulation in January, February or March **(Khidas, 1990; 1998)**. After a gestation period of 57 to 63 days, the female gives birth to 6-8 young **(Le Berre, 1990).**

According to (**Livet and Roeder, 1987),** the male genet has no typical seasonal sexual cycle, and the testes maintain a complete lineage throughout the year. The genet rut occurs in January-February, with a secondary rut in May-June. These periods are not fixed, however, and births can occur all year round. Gestation is 55 to 77 days with copulation lasting 3 to 5 minutes, one to two litters per year of up to young and late sexual maturity (2 years) **(Ahmim, 2019)** . Although little is known about population dynamics, including survival rates by age class and sex, the genet is one of the species with a slow reproductive strategy like the weasel and marten **(Croquet, 2005; Léger and Ruette ,2010).**

2.2.2.8 Social organization and behavior

Social organization in carnivores is based on three components **(Kapeller et al, 2013)**:

- Group size, composition and spatial/temporal distribution
- Reproduction methods
- The social structure that defines the interactions and kinship relationships between individuals in a group.

The golden wolf's social organization is extremely flexible. This is often structured around the breeding pair and the young from the previous litter. Depending demographic conditions and accessibility to resources, wolves can lead solitary or semi-gregarious lifestyles **(Moehlman and jhala, 2013).** Home range size is linked to the type of food making up an animal's diet **(Clutton-Brock and Harvey; 1978 in Judas, 1986)**. The Golden Wolf can travel very long distances in search of food and return its roost **(khidas,1998).**

The pups are raised by the couple, in which case hunting is done as a family, except when the female is nursing the pups, in which case hunting is done individually by the male. According to an ethnozoological survey we carried out in the past, retired hunters and nature observers believe that wolves tend to store extra food, often in burrows in the ground, away from other predators. For family sunbathing, the walk would be guided by the couple, in rocky areas, the highest spots are considered safer, for quiet naps, in places a little more exposed to human frequentation above all. The wolf digs a burrow and settles there out of sight, taking advantage of the sun's rays, especially during the cold season. It spends a little more time there, apart from sleeping during the day, and since it's nocturnal, it's preferable to take refuge in its own den. The predator's reflex is to choose a kind of windbreak, such as eroded soil that's not at the same level, the lower side representing a hidden bed on the surface of the ground, while the higher side intercepts the winds to allow the animal a clear hearing, on the other hand, at high altitudes it only needs an unobstructed view to be able to watch its back. Another type of burrow is known for food storage. The usefulness of the burrows can be distinguished by their dimensions, of which the one dedicated to napping is the widest, open at one end, with an average depth of 20 cm. The storage burrows are more like large holes, cylindrical in shape, with a small depth. According to **(khidas,1986).** The golden wolf exploits its living space differentially, i.e. the intensity of use of different parts of its home range is not uniform. Burrows dug by wolves should not be confused with those dug by wild boar, as wild boar are always on the lookout. The African golden wolf frequents urban environments much more frequently, and has become a visitor to the village of Seraidi every evening, often seen near garbage cans. It has developed an opportunistic behavior based squarely on feeding on garbage **(Belbel et al, 2022),** although local residents of Djbel l'Edough complain when it hunts their livestock (particularly chickens and lambs), it does not tend to be aggressive towards humans except when t- it is starving according to several testimonies.

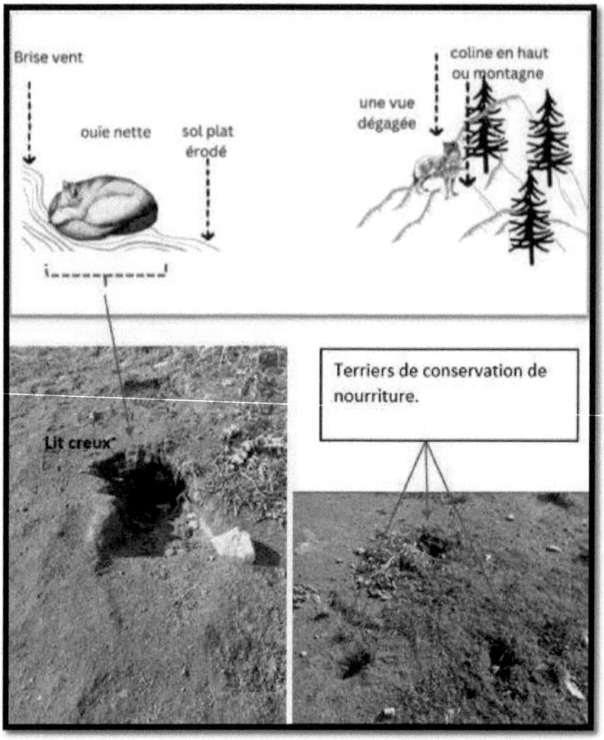

Figure12 : Diurnal behavior of the African golden wolf.

2.2.2.9 Diet

As a general rule, the African Golden Wolf is known to be opportunistic and omnivorous, and even often sets its sights on the carrion of wild boar often killed by hunters **(Photo 18)** and garbage **(Photo 19),** Confirmed on their part; this animal is an excellent mountain cleaner, this canid is also a skilled hntr Its diet includes birds, mammals, reptiles, fishinsects, fruit and vegetation **(Eddine et al, 2017; Boukheroufa** ***et al,*** **2020; Belbel et al, 2022)** in natural environments. It is opportunistic and feeds on garbage, noting the high proportion of vegetation in its diet. Arthropods are an important source of protein, and dominate its diet during the wet season, when small mammals are scarce **(Belbel et al, 2022).** Fruits play the role of energy food, including arbutus, forest fruits and bilberries **(BELBEL et al, 2022)**, whose seeds are visibly found in wolf droppings in the field, and are a very important part of its diet, especially during the dry season. The other very important part of the wolf's vegetation is the ***Ampelodesmos mauritanicus*** plant, commonly known in north-east Algeria as 'El diss', which acts as a digestive aid, either when it feeds on birds and mammals to easily rid itself of hair, or when it feeds on waste.

Photo 18: Carcass of wild boar *Sus scrofa* (© BELBEL.F)

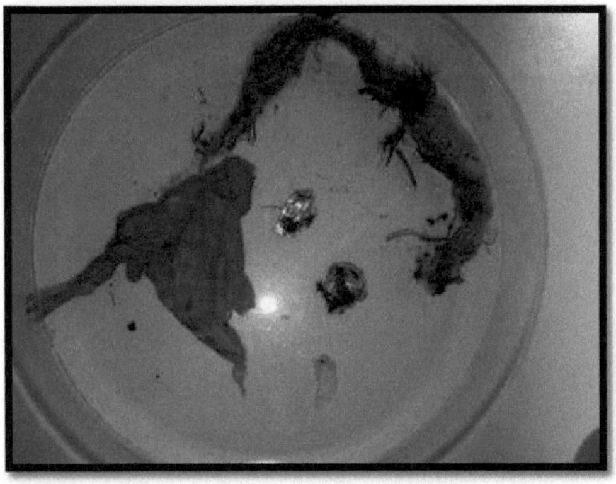

Photo 19: Waste category in the diet of the African golden wolf, (© Belbel).

2.2.2.10 Conservation status

African golden wolves are classified as "Least Concern" but declining on the IUCN Red List. **Durant et al (2011)** have also documented a significant long-term decline in golden wolf populations in the Serengeti. The reasons for this decline may be due to overkill by hunters and poaching, both present in the golden wolf's range, as well as retaliation by farmers for livestock predation. All this is aided by the growing stockpiles of automatic weapons in countries like Ethiopia. It is assumed that many golden wolves are also affected by predator control programs for other species, mainly through the consumption of poisoned carcasses. In addition, collisions with vehicles have been the source of death for at least fifty golden wolves in the Sahara Desert, which is likely to have greater implications as countries develop and roads become more complex and widespread. **(Brito, et al., 2009; Durant, et al., 2011; Eddine, et al., 2017; Moehlman and Hayssen, 2018; Yalden, et al., 1996).**

2.3 General methodology e

Our study was carried out in the Edough mountain range from December 2018 to May 2021. Temperature data for the Annaba region covering the study period were provided by the Office National de Meteorologie **(Table 2).**

Table2 : Meteorological data for the city of Annaba (2018- 2020). T: Mean annual temperature. TM: Mean annual maximum temperature. Tm: Average annual minimum temperature. PP: Total annual precipitation ta (Source: Office nationale de météorologie: www.meteo.dz)

Year	T	TM	Tm	Pp	V	RA	SN	TS	FG
2017	18	23.3	12.6	650.20	12.1	88	0	33	38
2018	18.1	23.4	13.1	618.22	11.9	118	0	50	25
2019	18	23.2	12.8	868.44	12	119	0	58	21
2020	18.2	23.7	12.7	655.79	12.8	95	0	28	19
2021	18,7	24,4	13. 3	619.8	13	89	0	31	29

3

2.3.1 Description of study sites

During the first year of sampling, we carried out an extensive campaign to collect genet and wolf droppings at six different stations, all located in the eastern part of the Edough mountain range: (1) Oued Semhout, (2)Ain Boukal (3)Berwaga, (4)Le Pont Romain, (5) El Chara, (6)Ain Barbar and (7) the Bouzizi dump.

We then focused on a site considered to be natural, located at Ain Boukal (Site 2) (36°93'03.25 "N 7°70 '95 .36 "E), a forested area far from settlements and comprising 4 successive tracks **(Figure 14).**

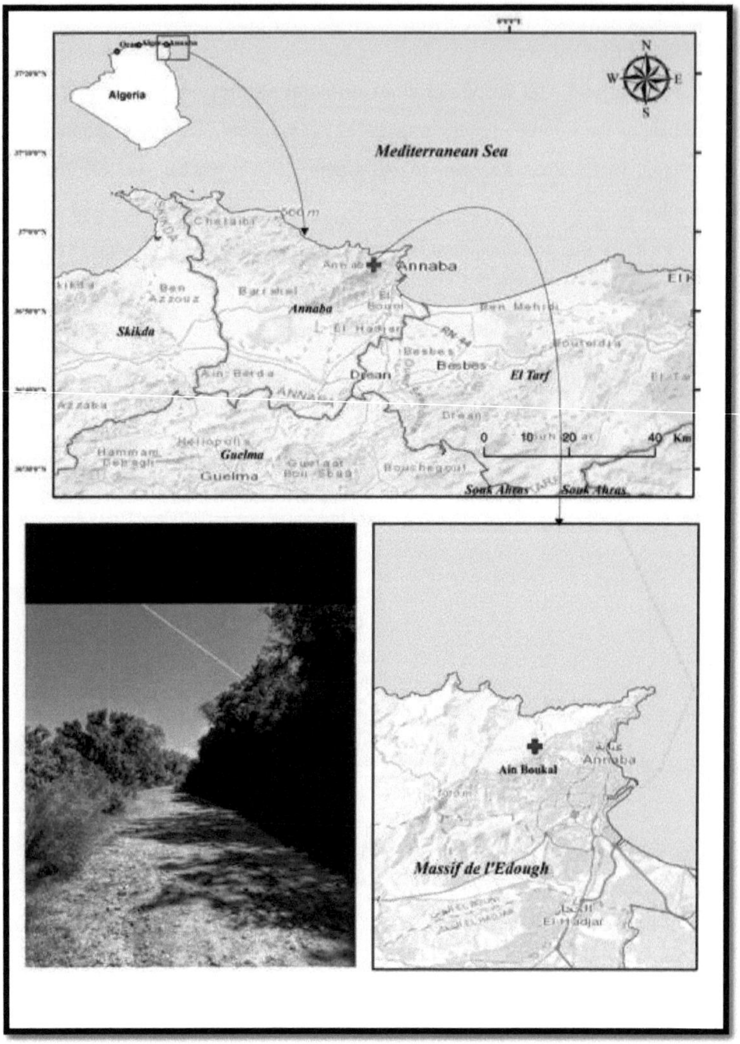

Figure13 : Presentation of sampling sites Ain Boukal natural site

2.3.2 Faeces sampling, identification and collection strategies

Following transects of 2.5 Km per site, and to distinguish between droppings, several criteria can be taken into consideration: size, odor, morphometric characteristics and place of deposition of droppings (**Macdonald, 1980; Bang and Dahlström, 1991**). The common genet (*Genetta genetta*), for example, is easily distinguished from other predators by droppings with an average diameter of 13.1 ± 15 mm, and an average length of 138.6 ± 23.4 mm, most often deposited on privileged high defecation sites, or rocky escarpments called crottiers **(Roeder, 1980; Lozé, 1984; Livet and Roeder, 1987). (Photo 10).** Other mammals, such as the African Golden Wolf and Jackal, deposit their droppings low to the ground, usually at the edge of paths **(photo 11).** Confusion between wolf and fox droppings, for example, is not common, since the latter's droppings are smaller, rarely contain bone fragments, and are composed mainly of the remains of small mammals, fruit residues and often contain insect fragments (**Eddine *et al.*, 2017**). The differentiation of African golden wolf dung from that of other carnivores living in the same area, such as the red fox (*Vulpes vulpes*), Egyptian mongoose (*Herpestes ichneumon*), wild cat (*Felis lybica*), and especially the domestic dog (*Canis lupus familiaris*) is based on specific characteristics such as odor, size, shape and location of deposits **(Macdonald, 1980; Bang and Dahlstrom, 1991)**. Our identification is also based on all evidence of the presence of the African golden wolf (howls, footprints, direct observations, observations by local residents), and on the fact that our natural study site is in forest, far from farms and residential areas, which explains the rarity of the presence of domestic dogs. As for the Bouzizi landfill, identification was facilitated by indications from local residents, the presence of footprints, and direct observation of the species near its burrow.

Photo10 : Common genet droppings on a rock (©Boukheroufa)

Photo11 ***: Scat of the African Golden Wolf (Canis anthus). ©Boukheroufa***

The crottier sites were chosen for their regular use. These sites were visited regularly, twice a month. As for golden wolf droppings, we followed a 2.5 transect along the trails to collect them. We also photographed the droppings and measured them (an essential parameter for determining the species). To do this, a ruler or some other object is placed close to the droppings to better appreciate their dimensions **(Table 3)**. Collecting faeces is a very delicate operation. Indeed, droppings are a source of contamination by pathogenic germs (echinococcosis, rabies...), so it's imperative to have no direct contact with the dropping. That's why it has to be placed in an airtight bag using two wooden chopsticks, picked up on the spot and then discarded.

Table3 . Dimensions of leashes (droppings, faeces, épreintes) of some Algerian Mammals. Source: Synthesized by Boukheroufa (2018)

Species	**Dimensions (mm)**	
	Length	**Diameter**
Porcupine (*Hystrix cristata*)	21 - 29	10 - 15
Weasel (*Mustela nivalis*)	25 - 50	2 - 3
Ichneumon mongoose (*Herpester ichneumon*)	45 - 80	8 - 12
Common genet (*Genetta genetta*)	100 - 250	15 - 20
African golden wolf (*Canis anthus*)	60 - 120	15 - 25
Red fox (*Vulpes vulpes*)	48 - 80	13 - 22
Wild rabbit (*Oryctolagus cuniculus*)	0,6 - 12	0,6 - 12
Red deer (*Cervus elaphus*)	18 - 25	12 - 20
Otter (*Lutra lutra*)	≈ 40 - 60	≈ 30 - 60

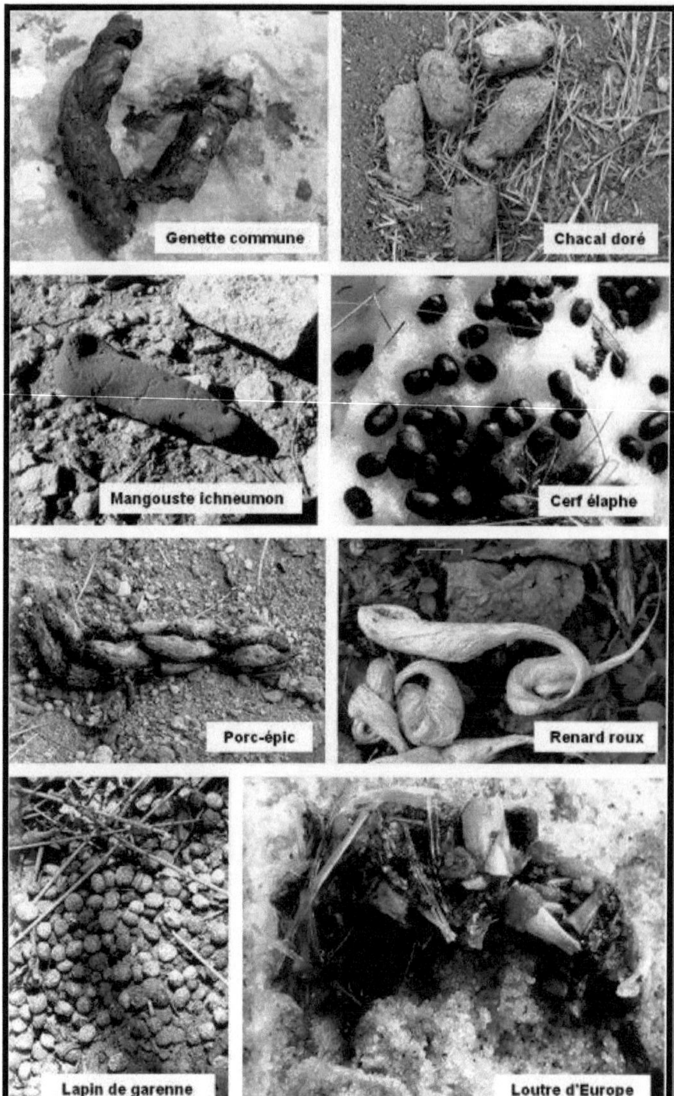

Photo12 : Some mammal droppings. Chazel and Chazel (2008).

2.3.3 Storage, treatment and sorting of faeces

Droppings collected in the field are placed in 70° alcohol for 48 hours **(Hamdine *et al*, 1993)** or in an oven at 80° C for 24 hours (present work) to eliminate any germs that may be present. Dehulling feces remains a delicate operation, as prey remains are highly fragmented, particularly bone debris, where the dentition is mostly isolated and unrelated to the mandible. For this purpose, we used the wet hulling technique. This was adopted for freshly collected droppings, where prey remains are condensed and compacted.

The principle of this technique is based on the separation of the elements contained in a pellet immersed in water. The technique comprises four stages **(Figure 17)**:

1st step: maceration of the droppings in a flask for 48 hours, to soften them. At the end of this stage, 5 ml of the maceration medium is collected, after shaking, for subsequent detection of earthworm bristles.

2$^{(eme)}$ step: dilaceration of the droppings using forceps and a lanceolate needle.

3eme step: The droppings are washed in hot water and detergent over a 0.25 mm diameter mesh sieve to remove grease and impurities, then dried for 24 to 36 hours at a temperature of 50°C.

4th step: The various components (feathers, hair, bones, teeth, invertebrate or plant debris) are sorted into labelled Petri dishes, categorized and weighed.

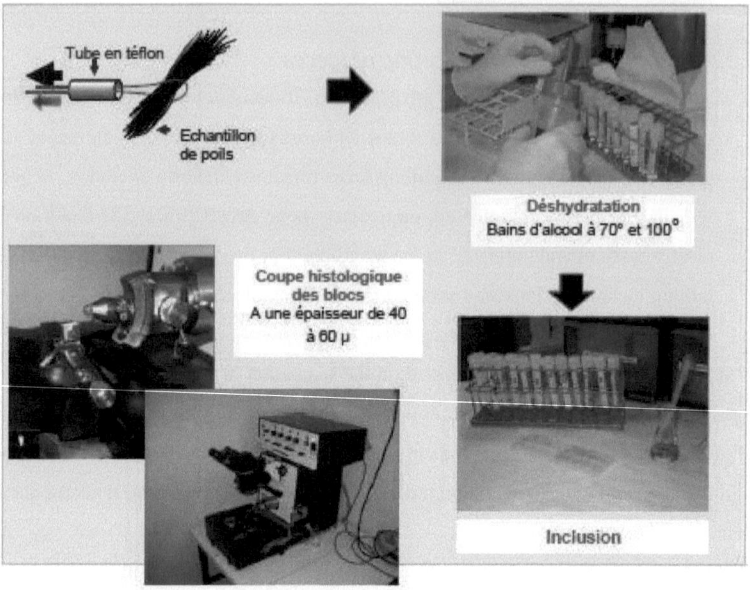

Figure14 : Histological protocol for dissecting small mammal hairs ***(©Boukheroufa)***

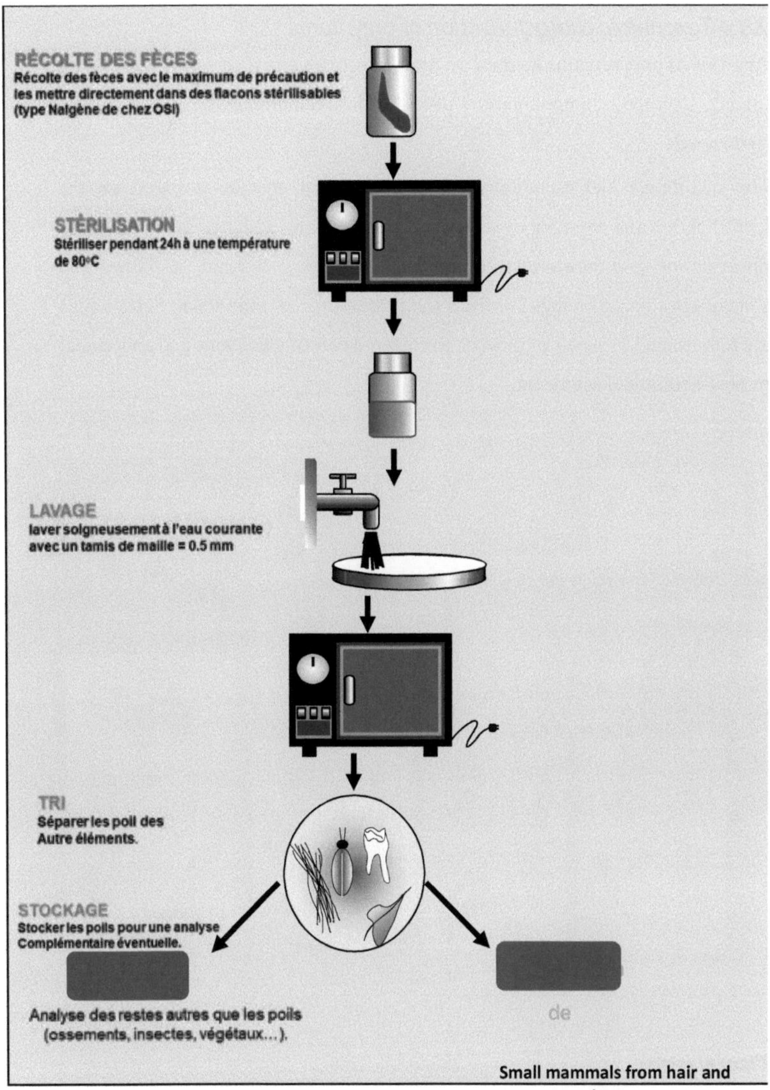

Figure15 : Carnivore faeces analysis protocol (Quéré, 1993 supplemented by Boukheroufa, 2005)

2.3.4 Taxonomic characterization of prey items

Analysis of prey remains is based on food categories and their respective classes (plants, arthropods, amphibians, re ptiles, birds and mammals).

Arthropods

The qualitative and quantitative determination of species is based on the identification and counting of anatomical parts making up individuals. Thus, the presence of a pronostomum (abdominal segments), a head, a thorax and sometimes a pair of wings confirms the presence of an individual. Scorpionids are represented by a pair of pincers, a stinger, a pair of chelicerae and abdominal or post-abdominal segments.

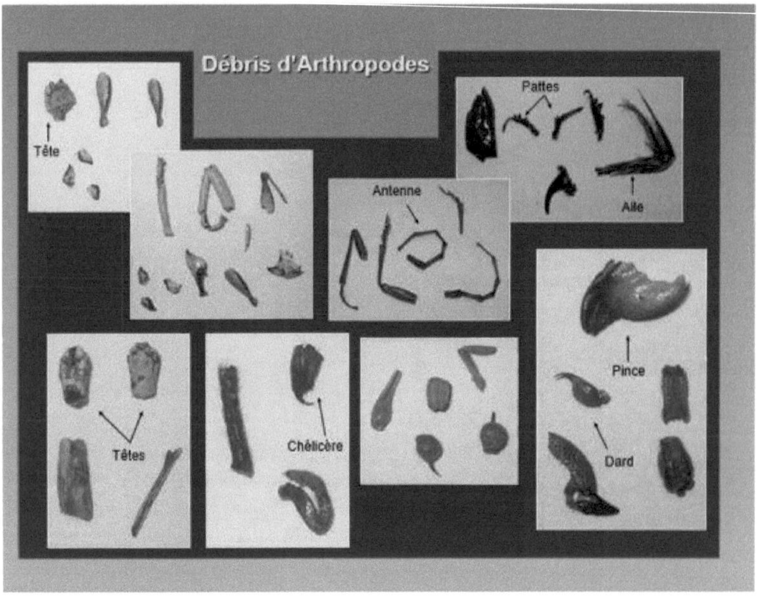

Plant elements

The fruit seeds and other plants found in the faeces are determined by comparison with the herbarium on the study site.

Amphibians and Reptiles

Amphibians and reptiles could not be identified due to a lack of identification keys. However, the presence of p legs or jaws in faeces could testify to their presence.

The Birds

Avian presence is indicated in feces by the presence of feather or down debris, sometimes legs, nails or a beak.

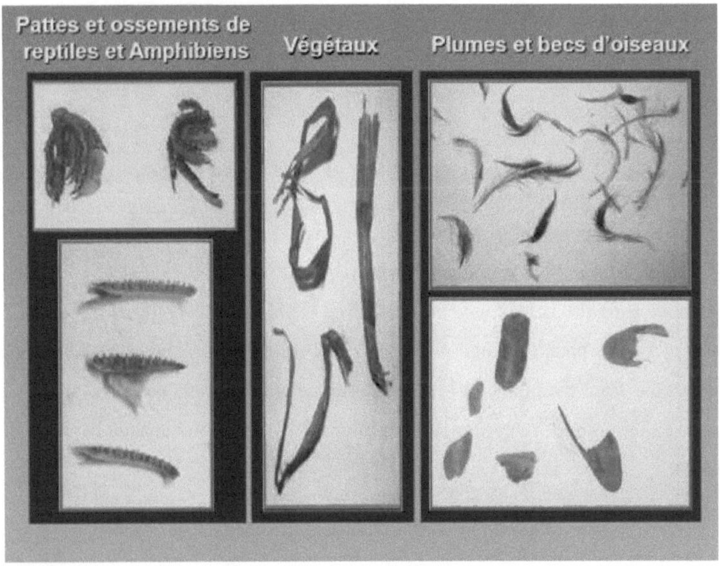

Mammals

Determination and enumeration of prey in this class is based on examination of the structure of recovered hairs, dentition characteristics and cranial biometry **(S aint - Girons and Petter, 1965).**

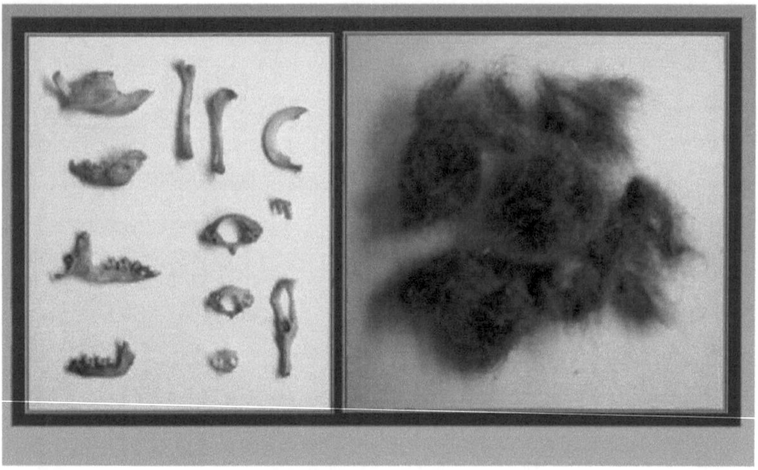

2.3.5. Hair analysis of small mammals

Examination of the hairs found in the faeces enabled us to confirm the data provided by the bone remains. A hair is made up of three cylindrical layers, interlocking from the outside to the inside: cuticle, cortex and medulla, all of which are keratinized. According to **Tupinier (1973)**, chiropteran hairs have no medulla.

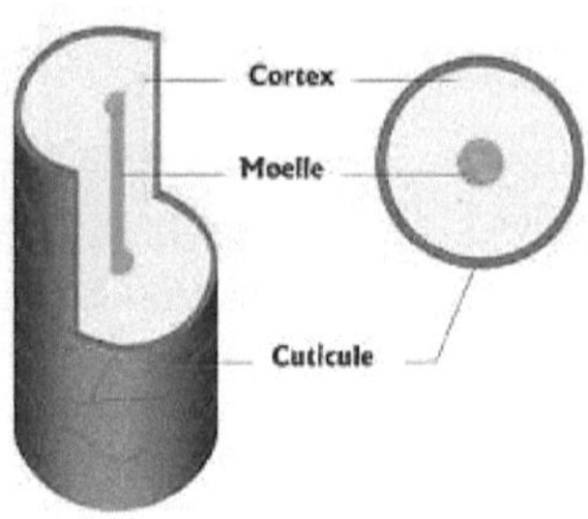

Figure16 : Diagram of a hair (Modified by BELBEL.F).

Examination of the hair found in the faeces enabled us to confirm the data provided by the bone remains. According to **Sari, A., & Arpacik, A. (2018)**, hair is one of the criteria that differentiates mammals from other animals. Three types of hair can be found in feces **(Figure 20)**: **Stuffing hairs** or underhairs, or down are fine, short and dense and much more numerous, play a thermal insulation role. - Kemp" **hairs**, along with undercoat hairs, make up most of the Micromammal coat. Kemp hairs are long, thick, less numerous and highly pigmented. - **Vibrissae:** keratinous hairs found on almost all mammals (feathers on birds), are few in number. They are found mainly on the head, tail and legs. They are the only hairs that taper regularly from base to tip. They are sensory organs with a role to play.

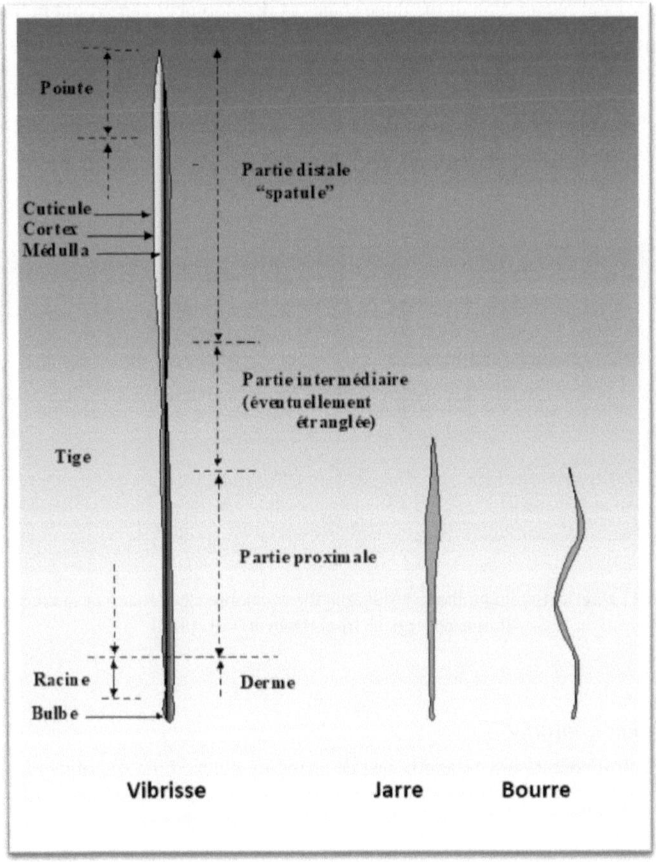

Figure17 : The different types of hair and diagram of a hair from **(Brunner et al. 1974; Faliu et al, 1979).**

2.3.6 Hair morphology

The hair shaft and tip are dead cells that do not evolve morphologically, whereas the cross-sections (A-A) **(Figure 21)** at the top represent the same stage of hair development. During our study, hairs were collected randomly from predator feces, so cross-sections were made at different stages of development, so we resorted to morphological recognition of the three stages of development carried out in the study of identification guides.

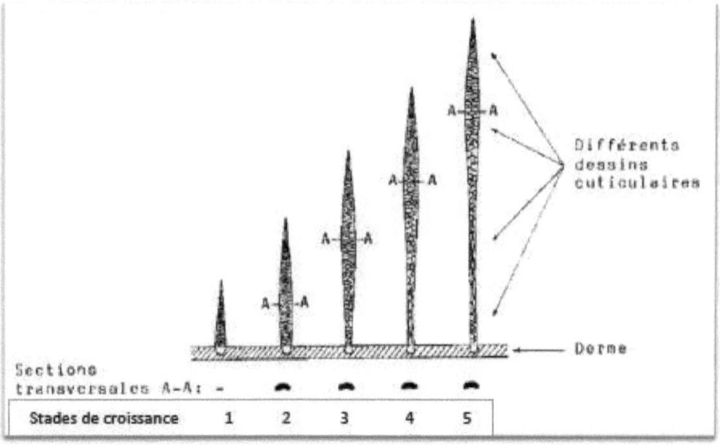

Figure18 : Hair morphology: the five stages in the development of a jar hair in the brown rat Rattus norvegicus from **(Debrotet al, 1982).**

2.3.7 Hair anatomy

Changes from one family to another while changing the medulla and other elements:

- Medulla
- Cortex and cuticle
- Cuticle

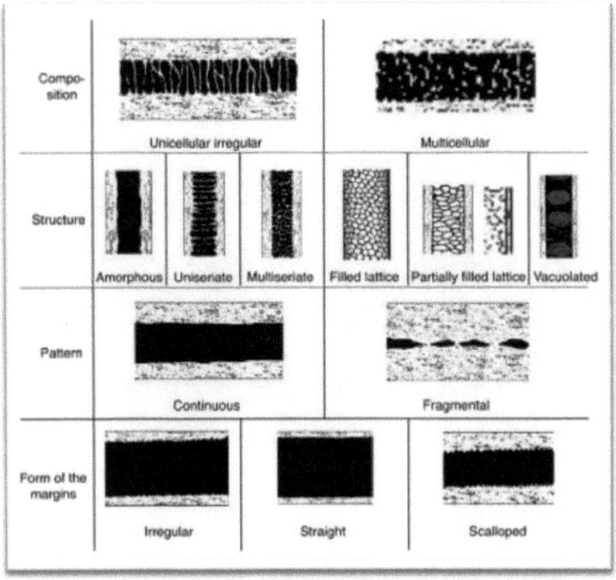

Figure19 : identification of angulates based on the three components of a hair (side view and profile) **(DeMarinis,A & Asprea,A, 2006).**

One of the keys to hair identification is the external morphology of the cuticle, i.e. the appearance of the hair, known as the imprint. This can be done in different ways **(DeMarinis & Agnelli, 1993 ; De Marinis & Asprea, 2006 ; Debrot, 1982)** , these methods expose the morphology of each animal's hair, and even differ from one species to another.

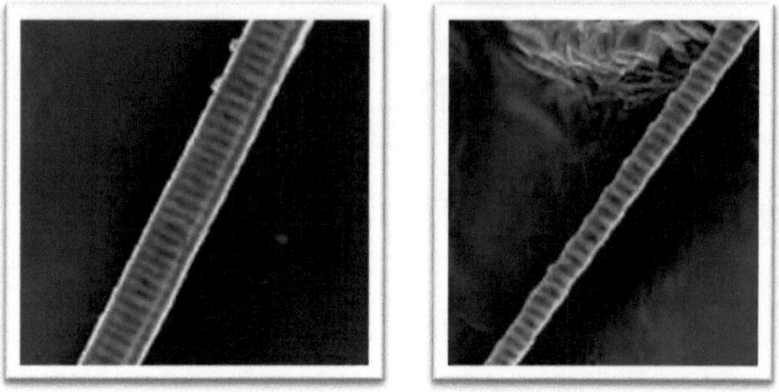

Figure20 :On the left, the cuticle of the shrew Blarina carolinensis; on the right, the cuticle of the shrew Cryptotis parva **(Debrot et al, 1982).**

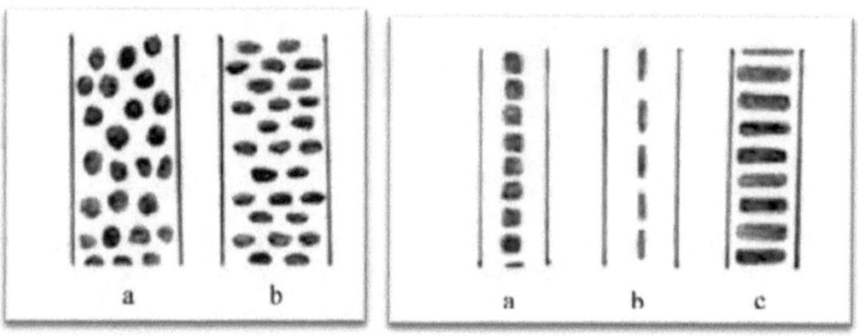

Figure21 : Types of compound medulla: left (a) oval and (b) flattened, right Types of simple medulla: (a) oval, (b) elongated, and (c) flattened (**Debrot et al, 1982).**

2.3.8 Cross-sections

By observing cross-sections of the hairs under a light microscope, we can see different forms, each belonging to a well-defined taxonomic genus:

- **Tetraconcave** or sometimes **triconcave** or **dumbbell** shapes indicate the presence of ***Apodemus*** field mice.
- **Reniform**, **monoconcave**, **biconcave**, **dumbbell** and **triconcave** shapes indicate the presence of ***Rattus*** rats and Mus mice.
- **Square** or **tetraconcave** shapes indicate the presence of ***Crocidura*** shrews.

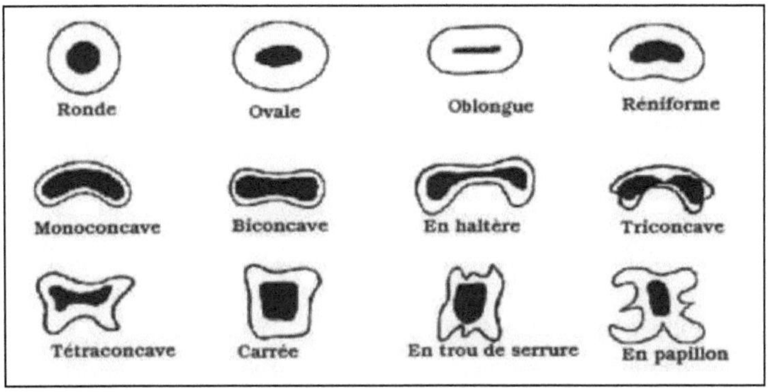

Figure22 : Different hair shapes observed under the microscope, cross-sections (*in* Damange, 1999).

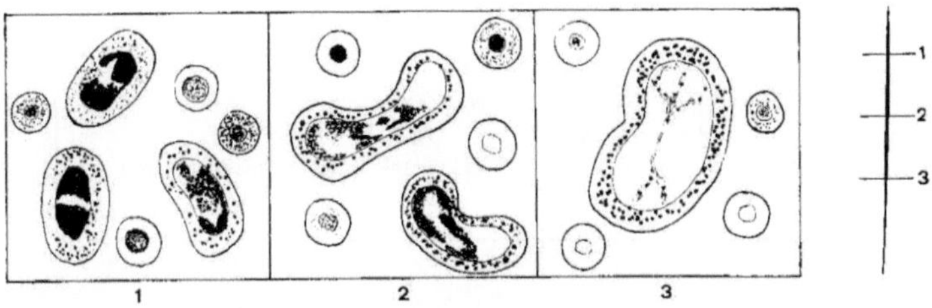

Figure23 : Differences in lateral hair structure within an individual Mus Musculus (after Debrot et al, 1982).

This diversity of cross-sectional shapes is the result of differentiations in the hair structure of this individual **(Debrot et al, 1982)**. However, it is important to point out that the only taxonomic identification reference based on hair is that of Debrot et al, and that it concerns the identification of European Mammals exclusively.

2.3.9 Bone analysis

In order to refine the taxonomic identification of the genera observed in the hair cross-sections, we moved on to examining and identifying the wear surfaces of the jugal teeth and mandibles with a binocular magnifying glass and comparing them with the determination keys **(Erome and Aulagnier, 1982; Barreau et al., 1991; Rolland, 2008; Boulanger, 2018).**

2.4. Data processing

We calculated and analyzed the population structure parameters of prey items (micromammals) consumed by the common genet and the African golden wolf.

2.4.1 Number of appearances (N.A)

It is defined as the number of times a food item or food category is found in all feces **(Lozé, 1984).**

2.4.2. Frequency of occurrence or Presence index (P.I. in %)

The presence index expresses the number of occurrences of each food item or food category out of the total number of faeces analysed:

$$\mathbf{I.\,P} = \frac{\mathbf{N.\,A}}{\mathbf{N_t}} \times \mathbf{100}$$

or: **N. A**: Number of
N_t: Total number of feces analyzed

2.4.3. Abundance (Ab)

Abundance is defined as the total population of a prey item (number of individuals) or food category (number of prey items).

2.4.4 Relative frequency or (relative abundance)

This is the Abundance of a prey item A in relation to the total Abundance of all prey items. It is expressed as a percentage.

$$\mathbf{F.\,R.} = \frac{\mathbf{Ab}}{\mathbf{Ab_t}} \times \mathbf{100}$$

or : **Ab** : Abundance of item A
Ab_t: Total abundance of all prey items.

2.4.5. Notion of biodiversity (H')

The diversity index expresses the structure of a population and the way individuals are distributed among various species **(Daget, 1979)**. We have adapted the SCHANNON and WEAVER index to calculate diet diversity:

$H' = - pi \log_2 pi$ or: **pi**: frequency of occurrence of each food item or food categories

This diversity is also known as intra-biotope diversity, and measures the level of complexity of the population (in this case, all the food items). The greater the number of food items, the closer their abundance and the greater the diversity of the diet.

The maximum diversity H' max is a function of the total number of food items **(*in* Virgos *et al.* 1999):**

$H\ max = \log_2 S$ or: **S**: Total number of food items ingested.

2.4.6. Equitability

Calculating equitability enables us to distinguish the extent to which the respective diversities (H') of the food items are close to their maximum diversity (H max), which corresponds to the stable equilibrium compatible with the environment. The equitability index is given by the following formula **(Virgos *et al.* 1999):**

$$E = \frac{H'}{H\ max} \times 100$$

or: **H'**: diversity

H max: diversity

2.4.7. The PIANKA index

We calculated the degree of overlap in the trophic niche of the two predators, based on the presence index (Pi) of the food categories, according to the Pianka index **(Pianka, 1973),** described by the following equation:

$$O_{jk} = \frac{\sum_{i}^{n} p_{ij} p_{ik}}{\sqrt{\sum_{i}^{n} p_{ij}^2 \sum_{i}^{n} p_{ik}^2}}$$

Where Ojk = measure of Pianka niche overlap between species j and k; Pij = proportion of food category i in the diet of species j; Pik = proportion of food category i in the diet of species k; n = total number of food categories. This index assumes that prey is equally available to all predators **(Reinthal, 1990).** Overlap values range from 0 (no overlap) to 1 (complete overlap).

2.6.7. Statistical analysis of data

Statistical analysis of the data was carried out using "Minitab 17" (*eds*, 2015) and R (*eds*, 2016) software. Results are expressed as means± standard error. We used Student's t-tests and analysis of variance (ANOVA) to compare means, the Chi-square test for categorical variables, and principal component analysis (PCA) to interpret the role of the different variables measured. The graphical representation of the data was based on histograms, sectors and rings produced in Microsoft Office Excel 2007.

RESULTS

III. RESULTS

At the end of this study, we collected a total of 403 droppings from all stations, 212 of which belonged to the common genet and 191 to the African golden wolf. This part, purely qualitative and based on taxonomic identification, was developed from the analysis of all the samples collected, i.e. the 403 droppings collected at different stations in the Edough mountain range.

3.1 Identifying micro mammals through their hair :

Observation of hair cross-sections under the light microscope, combined with the identification keys described *in* **Debrot et al. (1982),** enabled us to characterize at least four taxonomic genera: *Rattus, Mus, Apodemus and Crocidura.* We have also characterized forms that are frequent on section plans, but which we were unable to characterize, as they are not described in the Mammalian Hair Atlas alone (**Figures 22- 26**)

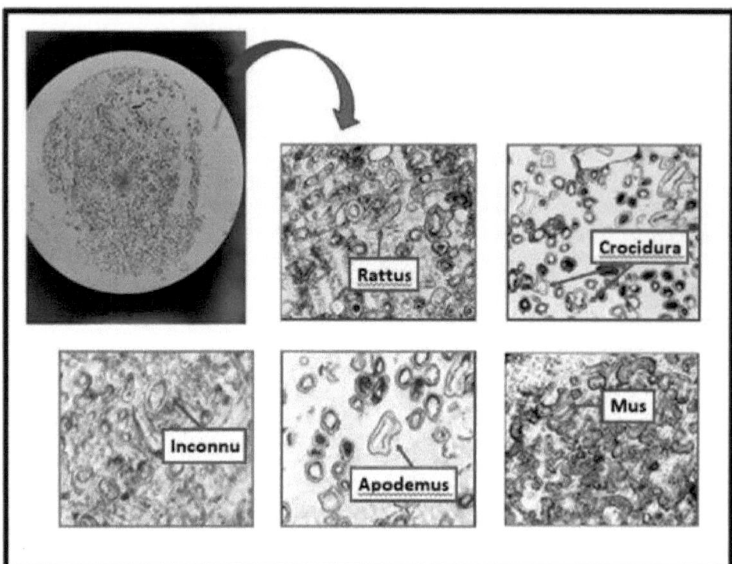

Figure 22: observation of hair cross-sections under the microscope

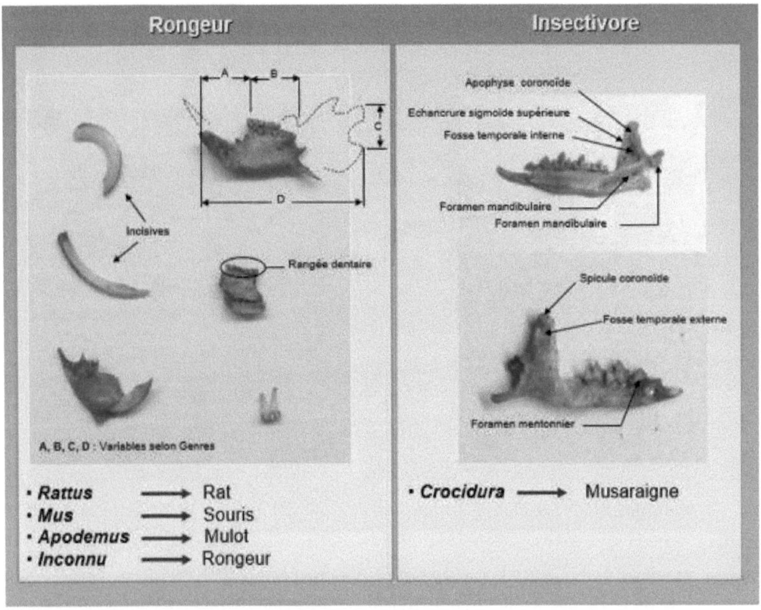

Figure 23: differences in bone structure between rodents and insectivores

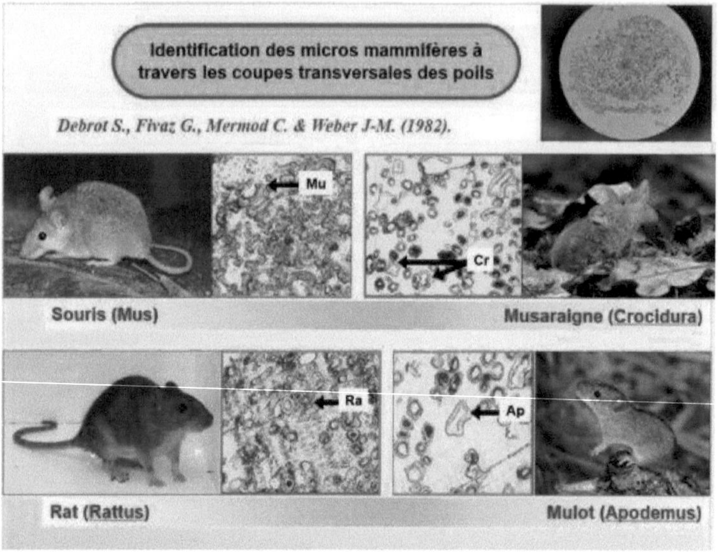

Figure 24: Identified taxonomic genera and unidentified forms from Gr (40*10) hair cross-sections (Cross-sections: Boukheroufa; Species photos: source google)

Of the 4 taxonomic genera formally identified (**photo 15**), three genera belong to the order **Rodents**: genus ***Apodemus,*** genus ***Rattus*** and genus ***Mus,*** and only one genus belongs to the order **Insectivores**, which is represented by ***Crocidura.*** Morphometric features that have enabled this characterization include:

- **Tetraconcave** or sometimes **triconcave** or **dumbbell** shapes indicate the presence of ***Apodemus*** field mice.
- **Reniform**, **monoconcave**, **biconcave**, **dumbbell** and **triconcave** shapes indicate the presence of ***Rattus*** rats and Mus mice.
- **Square** or **tetraconcave** shapes indicate the presence of ***Crocidura*** shrews.

This diversity of cross-sectional shapes is the result of structural differentiations in the hair of this individual **(Debrot et al, 1982)** (**figure 25**).

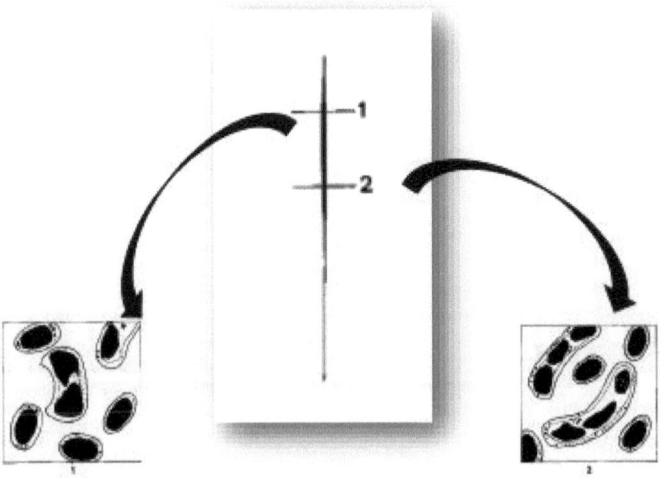

Figure24 : Differences in hair structure within an individual Mus Musculus **(after Debrot et al, 1982).**

3.1.1 Identification through bones

In order to refine the taxonomic identification of the genera observed in the hair cross-sections, we moved on to examining and identifying the wear surfaces of the jugal teeth and mandibles with a binocular magnifying glass and comparing them with the determination keys **(Erome and Aulagnier, 1982; Barreau et al., 1991; Rolland, 2008; Boulanger, 2018).**

We began by distinguishing the rodent order from the insectivore order on the basis of the general cranial and dental formula. Rodents have an ovoid skull, a rounded snout and a pair of developed incisors, separated from the other teeth (discontinuous dentition), enabling them to gnaw on the seeds and plants they feed on. Unlike rodents, insectivores have an oblong skull, with a pointed snout extended by a proboscis, their incisors are not as developed and are juxtaposed to the other teeth (continuous dentition) **(Rigaux & Dupasquier, 2012; Boulanger, 2018).**

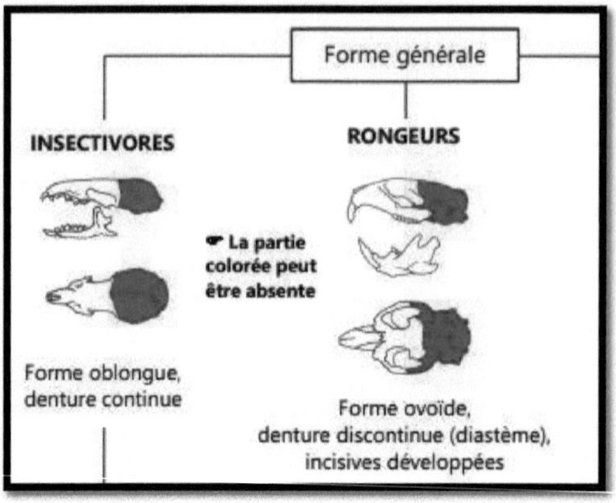

Figure25 : comparison of the general cranial shape of rodents and insectivores **(Boulanger, 2018).**

- **Insectivores :**

We sorted the bone remains, skull fragments and teeth that are characteristic of them **(Figure 28).** Each fragment was carefully observed under binoculars in order to accurately characterize the species.

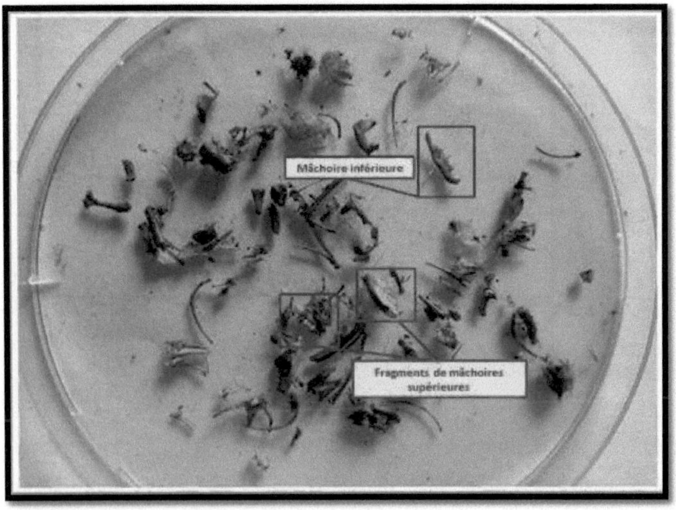

Figure26 : Identification of insectivore bones (Boukheroufa photo)

According to **(Rolland, 2008)** shrews are identified by the following criteria **(Figure 27. Figure 28)**:

1. Uninterrupted row of teeth (no diastema) (otherwise: rodents).
2. The median incisors are not separated from each other by a large concave enclave that rises into the palate (otherwise: Chiroptera).
3. The palate is longer than it is wide, with more than 4 teeth behind the first (and largest) canine: (Alternatively: *Mustela*, with 4 teeth).
4. Palate length less than 12 mm, width less than 8 mm. No zygomatic arches Shrews (otherwise: Mole, Hedgehog).

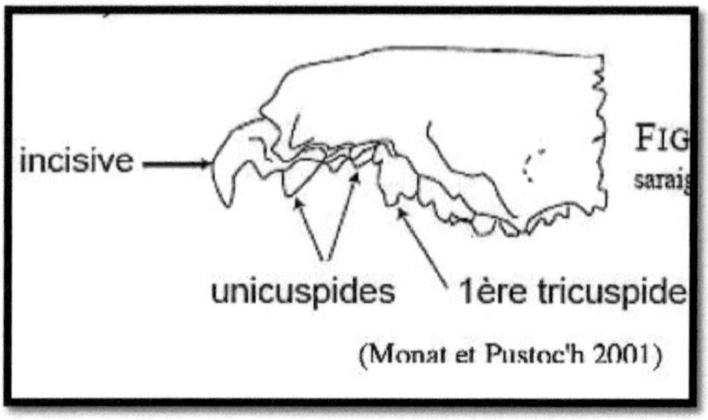

Figure27 : Shrew teeth **(Monat and Pustoch, 2001)**

The results of **Boukheroufa et al, (2009; 2020)**, **Belbel et al, 2022a,** combined with specimens found in Barber pots during crawling insect sampling **(Ouari and Bouhenniche, 2021)** confirm that this is *Crocidura russula.*

According to **(Rolland, 2008),** characterization is based on the number and color of unicupsids. This species has three unicupsids, all white compared with *Crossopes* (four unicupsids) and *Sorex* shrews (five unicupsids). **(Figure 30).**

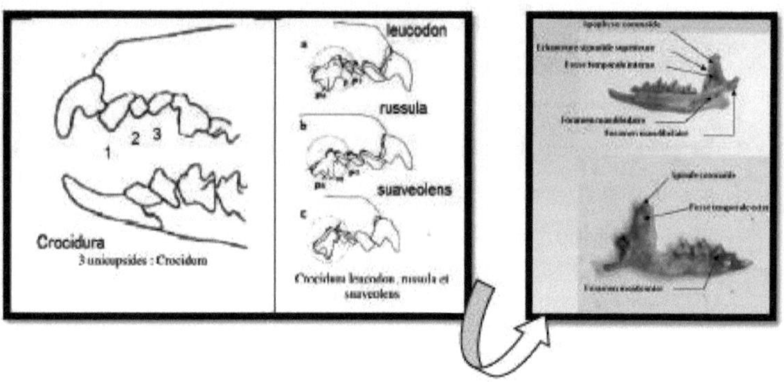

Figure28 : Crocidure mandible **(Aulagnier, 1982)**

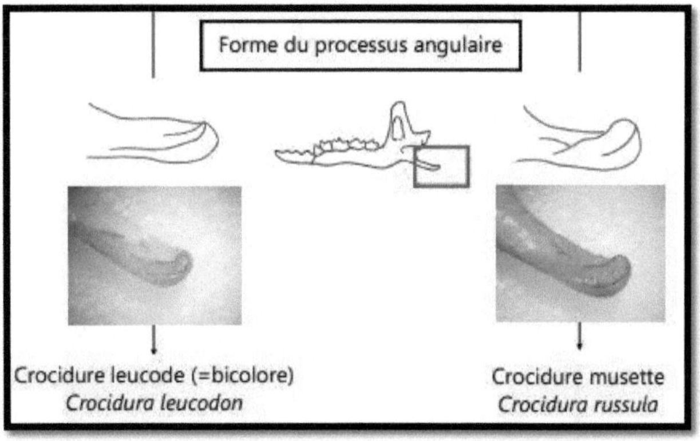

Figure29 : shape of the angular process in crocidura russula and leucodon **(Boulanger, 2018).**

- **Rodents:**

The bone remains corresponding to the Rodent Order (various incisors, mandible fragments, and the characteristic teeth of the Rodent Order) were carefully sorted and then examined under binoculars for taxonomic identification **(Figure 31; Figure 32; Figure 33)**

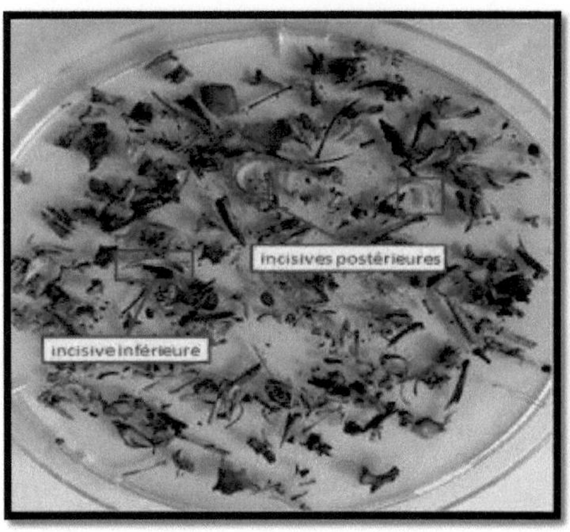

Figure30 : Identification of rodent incisors and molars

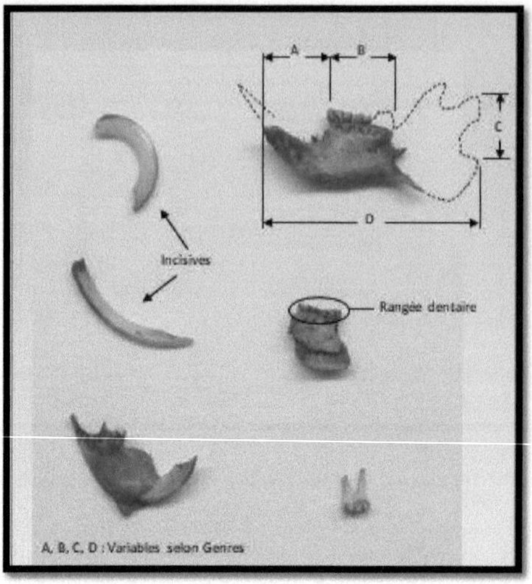

Figure31 : Identification of incisors and various fragments characteristic of rodents

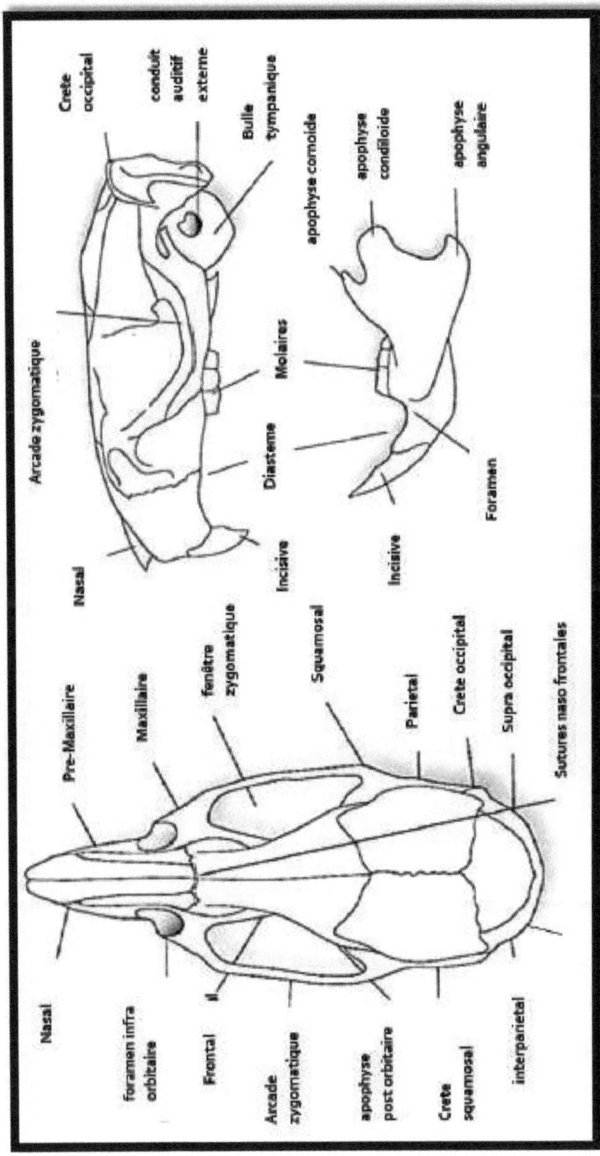

Figure32 : Morphology of a typical rodent skull (profile and top view) **(Couzi, 2011).**

The Figure shows the cranial structure and dentition of rodents:

- An interruption (bar or diastema) of more than two teeth in the dental rows, between the long incisors and molars.
- Canines and premolars are absent. On each half-jaw, an incisor is separated from a row of 3 molars.
- Ovoid skull, developed incisors.
- Murids have a tuberous molar wear surface, with rather low teeth and roots; they have three rounded tuberous molars, as in the case of field mice, rats and mice **(Rolland, 2008). (Figure 33).**

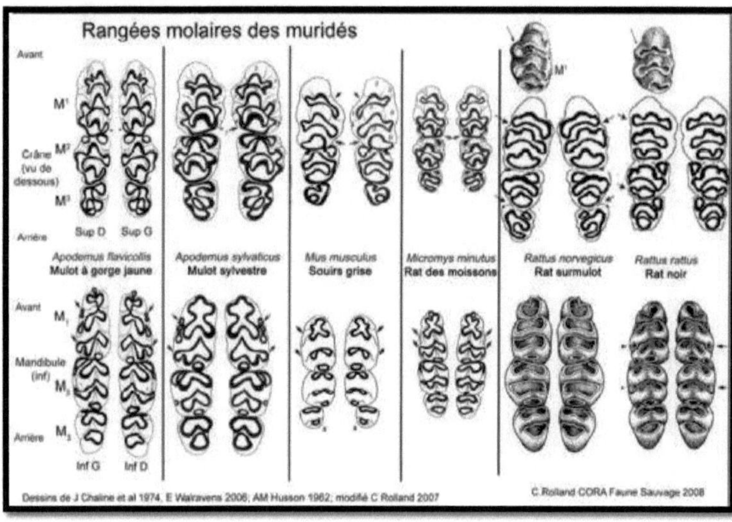

Figure33 : Molar rows in murids **(modified by Roland 2007)**

According to **Rolland (2008),** the Black Rat *(Rattus rattus)* is characterized by (**Figure 36**):

- Of the two lateral tubercles on the first lamella of the first molar, the inner is barely larger than the outer.

- These two lateral tubercles are quite distinct from the tubercles of the median lamella.
- There is no thickened rim around the front edge of the first molar lamella.

The same is described for field mice (Apodemus) and mice (Mus); they can be differentiated either by their molar rows (**Figure 37**) or by their incisors Another method also enables us to identify and differentiate between these three taxonomic genera belonging to the order Rodentia ; it is based on counting the roots of the first molar *(M^1) (*this molar is located furthest forward on the skull); by extracting the tooth with forceps and counting the holes in the bone **(Rolland, 2008; Couzi, 2011) (Figure 36):**

- **Three** roots confirm the identification of the **mice**.
- **Four** roots confirm the identification of **field mice**.
- **Five** roots confirm the identification of **Rats.**

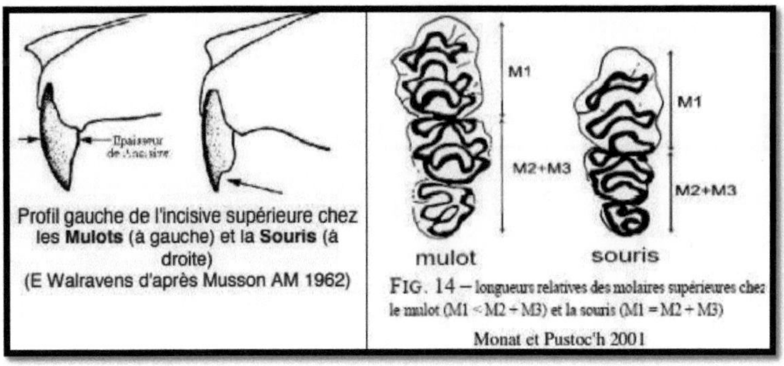

Figure34 : Comparison between field mice and mice

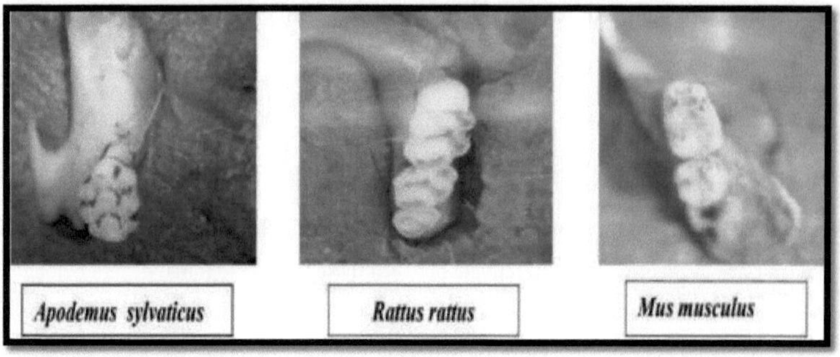

Figure35 : Identification of rodent species through **molar** rows **(In Boukheroufa, 2018).**

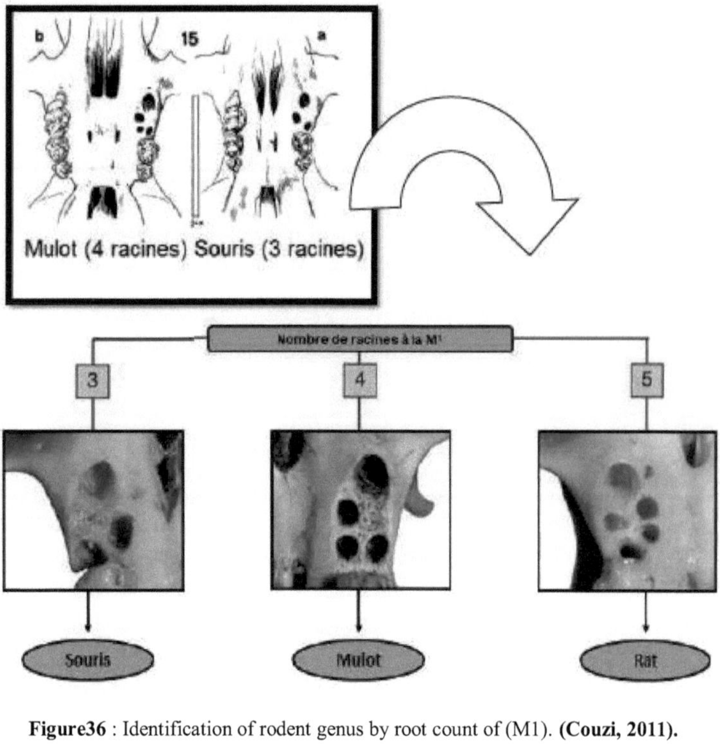

Figure36 : Identification of rodent genus by root count of (M1). **(Couzi, 2011).**

3.2.2 Overall analysis of the diet of the two predators in natural environments :

Analysis of the results reveals that the golden wolf consumes Fruits more frequently (present in 50% of excrement) than the other categories, followed by arthropods and Large and Micro Mammals (between 20 and 40%). Of course, this frequency trend needs to be analysed, as the energy value of this food intake needs to be considered in terms of biomass, where Large and Micro Mammals are larger prey that are much more satiating than Fruits. Waste and digestion plants (VD), in this case Diss (*Ampelodesma mauritanica*), are found in 25% of the excrement collected. The least regularly consumed category is Birds, with a frequency of occurrence of 12.5%. The common genet most frequently consumes small mammals (present in 88% of feces collected), followed by arthropods and birds (75% and 63% respectively). Fruits are found in 38% of feces. The genet also consumes the same amount of digestive plant matter as the golden wolf (25%). Waste and the reptile and amphibian category account for 13% of occurrence frequencies.

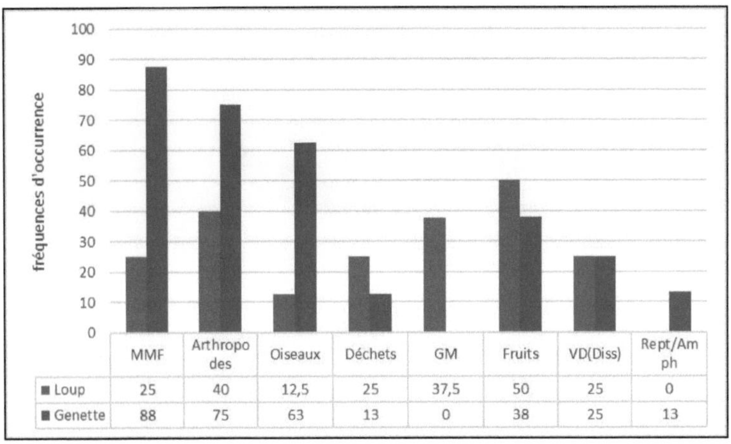

Figure37 : frequencies of occurrence of prey items consumed by the two predators.

Converting the frequency of occurrence into percentages has enabled us to characterize the diets of the two predators and highlight the share of each category. Fruit accounts for almost a quarter of the golden wolf's diet, while the common genet consumes small mammals and arthropods (over 50% of the diet). The African golden wolf is, in theory, more opportunistic than the genet, consuming resources of human origin, which are easier to acquire, with a proportion of 25%.

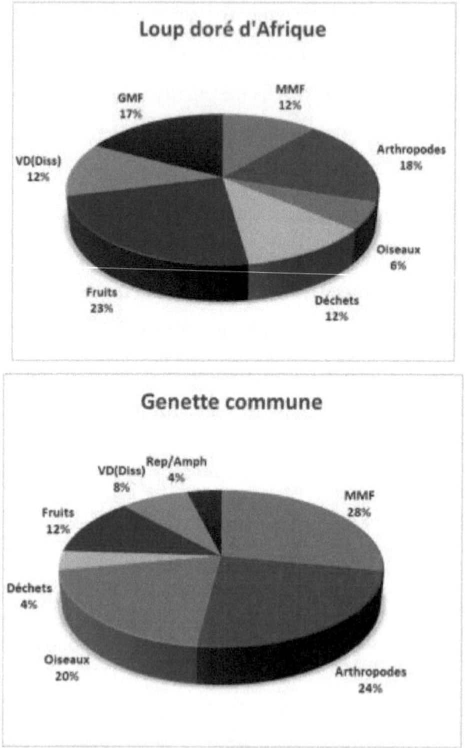

Figure38 : Analysis of overall diet composition in common genets and African golden wolves

Comparison of the trophic niches of the two species by calculating the Pianka overlap index reveals a value of 0.94, indicating almost complete overlap .

Calculation of the Pielou index reveals that the diet of the golden wolf is more diversified than that of the common genet (H'_{Wolf}= 2.70 / $H'_{(Genet)}$= 2.50) This result is confirmed by the permutation test with a very highly significant difference (*p= 0.0001*).

Measuring the frequencies of occurrence of the different food categories using Pielou's uniformity index shows, for the Wolf, a value closer to the maximum value of 1, which means that the distribution of food categories is more homogeneous compared to that of the Genet ($J'_{Wolf} = 0.96$ / $J'_{(Genet)} = 0.89$).

Comparative analysis of the frequencies of occurrence of micro-mammal prey in the two predators reveals a dominance of *Apodemus sylvaticus*, which is unquestionably the preferred prey of this predator in both the dry and wet seasons. The genet also consumes the genus *Rattus* at 35%, the genus Mus at 14% and the genus *Crocidura* at 3%.

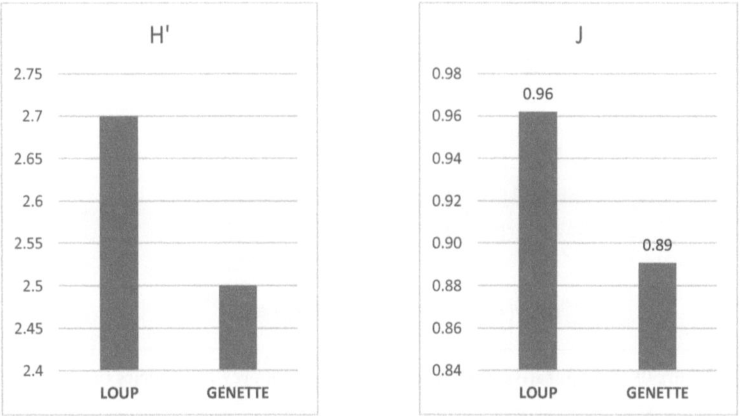

Figure39 : Pielou index calculation

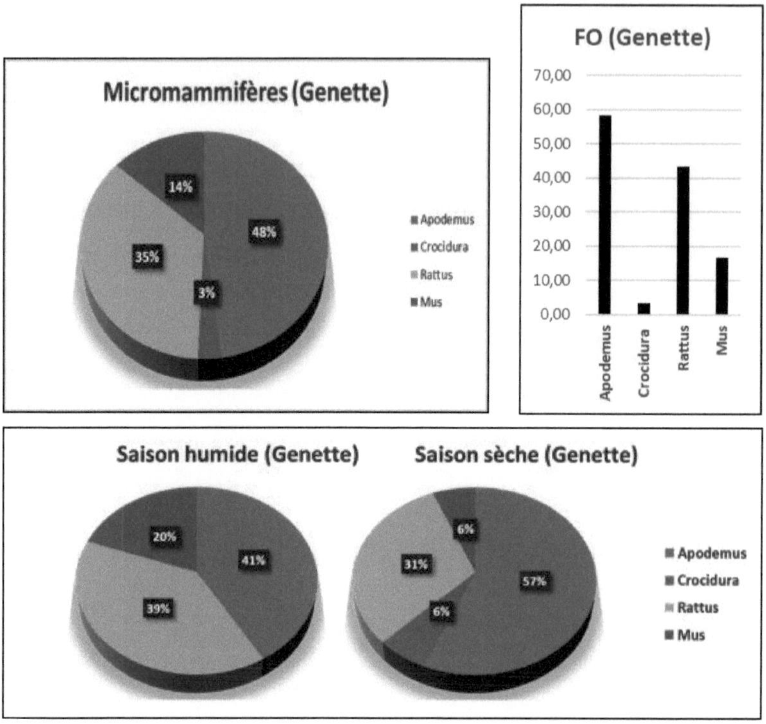

Figure40 : Comparative analysis of micro-mammal prey frequencies in genets.

In the African golden wolf, the diet is generally the same throughout the annual cycle, with a high degree of ecological elasticity and opportunism. The preferred micro-mammal prey in natural environments is the same as that of the common genet, namely *Apodemus sylvaticus*.

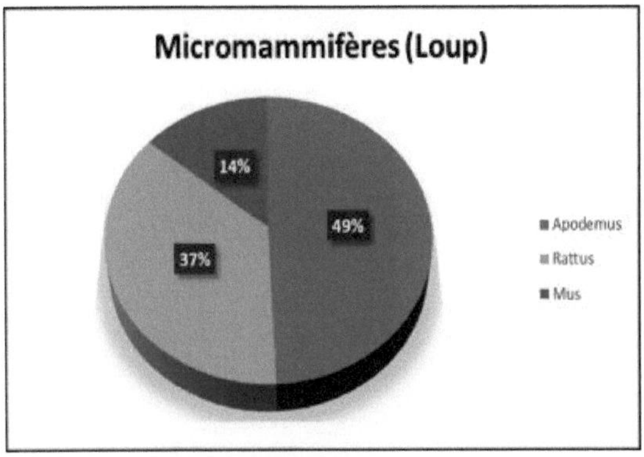

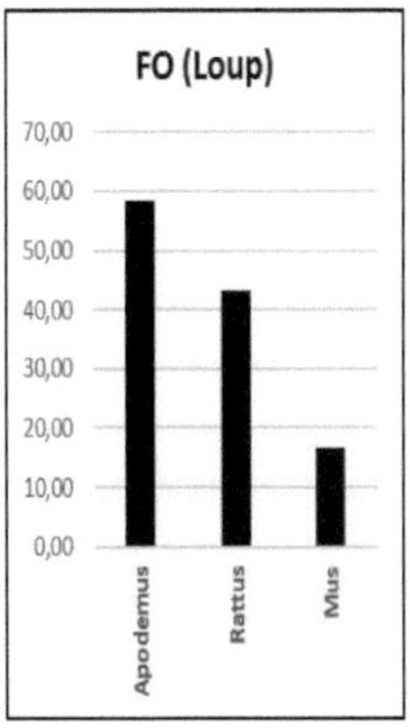

Figure41 : Comparative analysis of micro-mammal prey frequencies in wolves.

DISCUSSION GENERALE

IV. General discussion

Which small mammals do the two predators eat?

At the end of this preliminary taxonomic identification work, we were able to confirm the species consumed by the two predators: :

- Woodchat *(Apodemus sylvaticus).*
- Black rat *(Rattus rattus).*
- Grey mouse *(Mus musculus).*
- *Shrew (Crocidura russula).*

For each species identified using the combined keys derived from observation of histological sections and those derived from binocular observation of bones, we have produced a detailed description **(Mistrot, 2000; Oppliger, 2008; Ahmim, 2019; Zeribi, 2021).**

Apodemus sylvaticus (field mice) is a small rodent belonging to the Muridae family. They are fast-moving, good runners and jumpers, with long tails and well-developed ears. Their appearance is reminiscent of the house mouse. In the wild, field mice feed on seeds, small fruits and insects. In some countries, these mice are among the species most likely to carry the hantavirus. Other species in the genus include the striped field mouse (*Apodemus agrarius*) and the collared field mouse *(Apodemus flavicollis).*

Rattus rattus (rats) is a species belonging to a genus of large Muridae native to Asia, of which two species have colonized Europe and the rest of the world: the black rat (Rattus rattus) and the brown or Norway rat (Rattus norvegicus). They are commensals of man and live exclusively on human subsidies.

The genus *Mus* (mouse) is a genus of rodents in the Muridae family. Among the species belonging to this genus are Mus Musculus (grey mouse or house mouse) and *Mus Spretus* (wild mouse). Taxonomic identification has enabled us to characterize the *Mus Musculus* species. These small mammals resemble field mice, genus *Apodemus*. Only details of the skull allow us to differentiate them, as well as variations in dentition. They have a pointed snout, small eyes and large ears. They are omnivorous, but prefer cereals. Like rats, these mice originated in Asia. They have colonized the whole world, thanks to human activity. They can be found wherever humans live, in cities and on companionships.

The genus ***Crocidura*** is a group of insectivores belonging to the Soricidae family. This genus of shrews, with a pointed snout and extended white-toothed proboscis, includes a large number of species known as shrews or crocidures, such as the shrew (*Crocidura russula*) and the garden shrew *(Crocidura suaveolens).* In our study, we were able to characterize the *Crocidura russula* species. The Mediterranean zone, with its maquis, heather and good climatic conditions, is a good place for this shrew. Elsewhere, it frequents a wide range of biotopes, such as forests, meadows, edges, hedges and riverbanks, provided there is sufficient vegetation. It also likes dry-stone buildings **(Fons, 1984a; Genoud and Hutterer, 1990; Mardonald and Barrett, 1995; Lugon-Moulin, 2003; Oppliger, 2008).**

Photo13 : Shrews caught in the Edough mountain range (© BELBEL.F)

In addition to the species identified with certainty, we found unidentified (unknown) shapes on hair cross-sections, such as **round (Sp 1)**, **oval (Sp 2)** and **oblong (Sp 3) (Figure 25).** These shapesfound in many samples, may indicate the presence of other taxonomic genera specific to the Mediterranean region, and which do not appear in the European Mammal Hair Atlas. Once again, it's clear that Mediterranean environments are conducive to a high level of biodiversity, thanks to the mosaic of habitats they offer;This is the case, for example, of the Barbary Striped Rat of the genus *Lemniscomys*, which does not appear in Debrot's atlas, but the existence of this rodent has been confirmed in northern Algeria (**Bensidhoum, 2010; Mallil, 2012; Ikhlef and Boughedda, 2018)** and Morocco **(Denys et al, 2015).** The Field Gerbil, another rodent of the genus *Dipodillus* found in Morocco and the Sahara, is not also included in Debrot's catalog, but their characterization has been established on the basis of hairs and especially bones **(Denys et al, 2015)**. The ferret is a rodent of the *Eliomys* genus that has also been identified in northern Algeria **(Bensidhoum, 2010; Mallil, 2012)**

For insectivores, we were able to identify *Crocidura russula* with certainty. **Ahmim (2019)** reported the presence of 5 shrew species in Algeria, namely: *Crocidura russula, Crocidura whitakeri, Crocidura pachyura Crocidura cossyrensis, and Suncus etruscus,* Cros, Other insectivores have been described in Algeria, but have not been identified as prey item of the common genet, this is the case of the Pachyure of the *Suncus* genus in northern Algeria **(Bensidhoum, 2010; Mallil, 2012; Ikhlef and Boughedda, 2018).**

Of the two predators, which is the better sampler of micro-mammalian biodiversity???

Overall, our results reveal highly diversified diets between the two predators, a diversity that can be explained by the heterogeneity of Mediterranean habitats, which provide a wide range of prey **(Di Castri, 1973, Raven, 1973)**. With 6 food categories (large mammals, small mammals of the genus *Apodemus*, birds, arthropods, fruit and waste), the African golden wolf is more of a generalist, confirming the findings of several authors across its range, who describe a wide trophic range made up of reptiles, birds, rodents, mammals of different sizes, large numbers of insects and their larvae, it may even occasionally kill young gazelles or large prey, mainly wild boar and sheep **(Lodé et al., 1991; Amroun et al., 2006; Eddine et al., 2017; Karssene et al., 2019)**. In Senegal, the golden wolf has been observed attacking flocks of lambs **(Gaubert et al., 2012)**. In our case, the only small mammal consumed is *Apodemus sylvaticus* and the only large mammal consumed is wild boar, whose density is high in the region **(Boumendjel et al., 2016)**. The African golden wolf probably feeds on the corpses of hunted wild boar, since our study period coincides with the wild boar hunting season in the region **(Zemiti, 2012). Eddine et al. (2017)** showed that wild boars were the most important prey in the Tlemcen hunting reserve, and that adult wild boars were probably consumed mainly as prey, due to their tusks and aggressive behavior **(Jedrzejewska and Jedrzejewski, 1998; Moehlman and Jhala, 2013)**.

The common genet is generally classified by many researchers between specialist species, such as the European otter (*Lutra lutra* (Linnaeus 1758)), and generalist species, such as the European badger (*Meles meles* (Linnaeus 1758)) **(Delibes et al, 1989; Lodé et al., 1991; Ruiz-Olmo and Lopez-Martin, 1993; Carvhallo and Gomes, 2003; San- chez et al., 2008; Le Jaques and Lodé, 1989; Camp, 2012; Amroun et al., 2014; Torre et al., 2013, 2015)**. Our results also confirm this intermediate position of the genet, which consumes only 4 categories (micromammals, birds, arthropods and fruits). The African golden wolf and the genet are generalist species, and this characteristic gives them the opportunity to adopt different feeding strategies and thus avoid competitive pressure, particularly for genets, which are the most vulnerable to different forms of interspecific competition **(Caro and Stoner, 2003)**. Niche overlap refers to the partial or complete sharing of resources or other ecological factors (predators, feeding area, soil type, etc.) by two or more species **(Cornell, 2011)**.

In winter, diets consist mainly of fruit and large mammals for the African golden wolf, and small mammals and birds for the common genet. This compromise in the use of environmental resources has also been observed between these same species in Kabylie (Algeria) by **Amroun et al. (2006)**. This phenomenon is explained by the size of the mammals consumed; the African golden wolf consumes large mammals that are inaccessible to the genet, and which represent a greater source of energy, such as cattle. Its trophic spectrum may also include small mammals. Competitive pressure can also be reduced by its tolerance of human activity, as it can feed on human waste **(Eddine et al., 2017; Yirga et al., 2017)**. This great ability to exploit all environmental resources, including those of anthropogenic origin and "easy to acquire", is less evident for the genet, which seems more vulnerable to any change in the environment, in particular to the fragmentation or deterioration of forest environments **(Gomes and Giraudoux, 1992; Gomes, 1993)**.

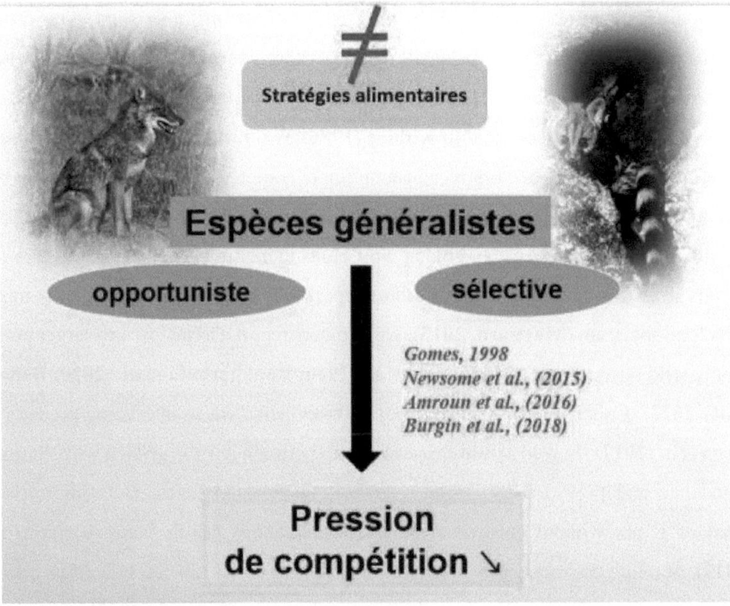

Figure 43: Explanatory diagram of the feeding strategies of the common genet and the timber wolf.

Does *Canis anthus* really regulate natural populations of small mammals?

The African golden wolf is a predator that exploits trophic resources according to their local availability and seasonality **(Amroun *et al.*, 2014. Eddine *et al*, 2017; Irzagh *et al.*, 2020)**. In this study, we examined the trophic opportunism of this predator, through its consumption of small mammals in the wild and in environments where trophic resources are abundant. The results obtained in the wild are comparable to those described by **Boukheroufa et al. (2020)**. They identified rodents in the species' diet and found a partial trophic niche overlap with another predator, the common genet, which is more selective towards small mammals. Thus, the wolf adjusts its diet by relying on wild boar carcasses, which are particularly abundant during prescribed hunting periods, in order to reduce the interspecific competition that could be exerted on small mammal populations **(Zemiti, 2012; Boumendjel *et al.*, 2016)**. In addition, wild boar carcasses are an easily accessible prey for *Canis anthus*, which does not require energetic effort to hunt, unlike small mammals. On the other hand, it turns out that the predator doesn't hunt the small mammals that swarm in the dump, but rather consumes the organic waste, enabling the animal to save its energy and allocate it to other vital functions. According to **Khidas (1986)**, two methods of foraging are used by the African golden wolf: firstly, consumption of resources not hunted, but rather by chance or local knowledge (in our case, wild boar carcasses, fruit, waste, etc.), and secondly, hunting. The African golden wolf hunts alone, in packs or groups, prey of various sizes **(Macdonald, 1983)**. Authors report that small mammals make up most of its food spectrum **(Maynard, 2015)**, while others report a trend towards larger prey, mainly wild boar and sheep **(Amroun *et al.*, 2016; Boukheroufa *et al.*, 2020; Belbel *et al.*, 2022)**. Concerning the presence of anthropogenic waste in feeding predators, **Yirga *et al* (2017)** showed a positive relationship between African golden wolf, human abundance and prey density depleted anthropogenic landscapes. But this trophic behavior is not without consequences for the predator's health status **Yirga *et al* (2017)**. Wildlife disturbance is one such impact, and could ultimately lead to major disturbances and affect the health and survival of species **(Xu et al., 2022).**

CONCLUSION AND OUTLOOK

V. CONCLUSION AND OUTLOOK

Population structure represents the main form of existence for all small mammals, and fully reflects the adaptive response to environmental change, including regional patterns of intra- and inter-specific variability. Micromammals are therefore key tools for understanding small-scale responses to environmental change, and are excellent bio-indicators of the quality of natural environments. This status is reinforced by their intermediate trophic position, depending on vegetation and arthropods for food, and interacting with other wild species through competition for resources or as a source of prey.

As part of this study, we carried out empirical taxonomic work, in order to provide specialists with a reference document detailing the methodology for processing and analyzing the bone remains and hairs of small mammals with a view to their taxonomic identification. Mastery of the latter will ultimately enable monitoring studies to be carried out, and improve knowledge of the diversity and structure of this population in a forest zone characterized by a succession of plant species conducive to a richness characteristic of Mediterranean environments.

Based on these results, it would be interesting to consider the following perspectives:

- Carry out a genetic analysis of hairs to characterize other micro-mammalian species not included in the European identification atlas.
- Carry out comparative studies in other ecosystems with different ecological characteristics, to gain a better appreciation of micro-mammalian diversity and dynamics.
- Characterize micro-mammalian diversity by studying the diet of other sympatric species such as red foxes, weasels, owls.
- Consolidate identification data using the direct method, by installing trapping devices (Shermann traps, Fosse traps, etc.).

REFERENCES BIBLIOGRAPHICA

VI. BIBLIOGRAPHICAL REFERENCES

1. Adamou-Djerbaoui, M., C. Denys, H. Chaba, M. M. Seid, Y. Djelaila, F. Labdelli, and M. S. Adamou. 2013. "É Tude Du R Égime a Limentaire D ' Un R Ongeur N Uisible (M Eriones Shawii D Uvernoy , 1842 , M Ammalia , R Odentia)." Lebanese Science Journal 14(1):15-32.
2. Adamou-Djerbaoui, Malika, Fatiha Labdelli, Yassine Djelaila, Karima Oulbachir, Mohamed Sofiane Adamou, and Christian Denys. 2015. "Inventaire Des Rongeurs Dans La Région de Tiaret (Algérie)." Travaux de l'Institut Scientifique No. 8:105-12.
3. Ahaggar, De and Laurie Marker. n.d. "inventaires de la faune du groupe d' interet sahelo saharien partie inventaires de la faune du groupe d ' interet sahelo-saharien partie 4 : Massif Central de l ' Ahaggar , Algérie Smithsonian." (December 2013).
4. Ahmim, M. (2019). The wild mammals of Algeria: Distribution and Conservation Biology. Les Éditions du Net.
5. Alden, P., Estes, R. D., Schlitter, D., & McBride, B. (1996). Field guide to African mammals. Houghton Mifflin Harcourt.
6. Alia, Zeid, Elamine Khachekhouch, Djilani Ghamem Amara, Nacereddine Tennech, Makhlouf Sekour, and Karim Souttou. 2020. "The Rodents Damages Caused on Arachis Hypogaea Andpisum Sativumcultures in the Souf Region (Algeria." PONTE International Scientific Researches Journal 762.
7. Amigues, S. (1999). The weasels of Tartessos. Anthropozoologica 29: 55-64.
8. Amroun M, Bensidhoum M, Delattre P, GaubertP (2014). Feeding habits of the common genet (Genetta genetta) in the area of Djurdjura, north of Algeria. Mammalia, 78, 35-43.
9. Amroun M, Giraudoux P, Delattre P. A (2006). Comparative study of the diets of two sympatric carnivores - the golden jackal (Canis aureus) and the common genet (Genetta genetta) - in Kabylie, Algeria. Mammalia,40, 247-254.
10. Anna Maria De Marinis, Alessandro Asprea "Hair identification key of wild and domestic ungulates from southern Europe," Wildlife Biology, 12(3), 305-320, (1 September 2006).
11. Auffray, Jean-Christophe, 1988 (Le commensalisme chez la souris domestique : origine, écologie et rôle dans l'évolution chromosomique de l'espèce). Doctoral thesis in Population and Ecosystem Biology, Université Montpellier II, 170p.
12. Aulagnier S, Haffiner P, Mitchelljones AJ,Moutou F, Zima J (2008). Guide of Mammals in Europe, North Africa and the Middle East. Delachaux-Niestlé, 271 p.
13. Aulagnier, S. (1992). The large carnivores of North Africa: Extinction and conservation. African Wildlife Research, 23(4), 157-167.
14. Aulagnier, S., & G. B. (2010). Mammals: Diversity and conservation. Muséum national d'Histoire naturelle.
15. Aulagnier. Stéphane. Zoogeography of the Mammals of Morocco (1992): from specific analysis to settlement typology on a regional scale. Diss. Montpellier 2.
16. Bang, P., & Dahlström, P. (1991). Animal tracks and signs. Oxford University Press.
17. Bang, P., & Dahlström, P. (2001 *). Animal Tracks and Signs*. Oxford University Press.
18. Barbault, R. (1994). Ecologie des populations et des écosystèmes. Masson.
19. Barbault, R. (1995). Biodiversity, global change, and the trophic network. Springer-Verlag.

20. Barbault, R., & Dajoz, R. (1995). The population-environment system: Functional and evolutionary perspectives. Ecology Review, 15(2), 80-95.
21. BAZIZ B, (2002). Bioecology and diet of some birds of prey in different localities in Algeria. Cas du Faucon crécerelle Falco tinnunculus Linné, 1758, de la Chouette effraie Tyto alba (Scopoli, 1759), de la Chouette hulotte Strix aluco Linné, 1758, de la Chouette chevêche Athene noctua (Scopoli, 1769), du Hibou moyen - duc Asio otus (Linné, 1758) et du Hibou grand - duc ascalaphe Bubo ascalaphus Savigny, 1809. Thesis, Doctorat d'Etat sci. agro, Inst. nati. agro, El Harrach, 499 p.
22. Bebba, Kaouther and Belkacem Baziz. 2011. "Micromammiferes Dans La Vallee D'Oued Righ." Biodiversité Faunnistique En Zones Arides et Semi-Arides 235-39.
23. Beddiaf, Rahma. 2012. "Etude Du Régime Alimentaire de Deux Rapaces: Le Hibou Ascalaphe Bubo Ascalaphus (Savigny, 1809) et La Chouette Chevêche Athene Noctua (Scopoli, 1769) Dans La Région de Djanet (Tassili n'Ajjer, Algérie)." Memoire Ing. Agro, Uni. Kasdi Merbah, Ouargla 103.
24. Belbel F, Boukheroufa M, Sakraoui Rym, Abdelli M, Benotmane C H, Henada R L I, Sakraoui F (2022). Does the african golden wolf Canis anthus, plays the role of natural micromammal populations in the Edough moutain range (northeastern Algeria)?. Uttar pradesh journal of zoology.43 (16) : 1 - 6.
25. Belbel, F., Boukheroufa, M., Benotmane, C., Sakraoui, R., Henada, L. I., & Sakraoui, F. (2022). Selection Strategy of Small Mammalian Preys by the Common Genet Genetta Genetta between Natural and Anthropized Environments in Edough Forest Massif (Northeastern Algeria), Journal of Bioresource Management, 9 (4): 9p
26. Bensidhoum M (2010). Spatial occupation strategies and trophic ecology of the Genette Genetta genetta Linne,1758 in the forest of Darna, Djurdjura oriental, Algeria. Tizi Ouzou, Magister thesis, departement of Biology, University Mouloud Mammeri of Tizi Ouzou, Algeria, 100 p.
27. Bernard, J., 1969. Mammals of Tunisia and neighboring regions. Bulletin de la Faculté d'Agronomie de Tunis, 24- 25 : 40-160.
28. Bitam, I., J. Rolain, T. Kernif, B. Baziz, P. Parola, and D. Raoult. 2008. "Bartonella Species Detected in Rodents and Hedgehogs from Algeria." Clinical Microbiology and Infection 15:102-3.
29. Bossière, J., & al. (1978). Geological features of the Edough Peninsula: A study of crystalline domes. Geological Society of Algeria Bulletin, 19(2), 201-215.
30. Boukheroufa .M, 2009. Ecologie alimentaire de la genetta commune Genetta genetta dans un ecosysteme forestier du parc national d'El Kala (Nord Est Algérien)Mésogée : bulletin du Musée d'histoire naturelle de Marseille. Marseille: Musée d'histoire naturelle de Marseille, 1986. Print.
31. Boukheroufa .M, 2018. Référentiel Méthodologique Algérien REFMETAL Mammifères.
32. Boukheroufa, M., Sakraoui, F., Belbel, F., & Sakraoui, R. (2020). Winter diet of the common genet, Genetta genetta (Carnivora, Viverridae), and the African golden wolf, Canis anthus (Carnivora, Canidae), in altitudinal locality of the Edough Forest (northeastern Algeria). Zoodiversity, 54(1), 67-74.
33. Boukheroufa. M. (2005). Ecologie alimentaire d ela genette commune Genetta genetta dans le parc national d'el Kala. Magister thesis. Université Badji Mokhtar Annaba. PP 85.
34. Boulanger, A (2018) ; Clé de détermination des proies des rapaces nocturnes des Hautsde-France (Key to determining the prey of nocturnal raptors in Hautsde-France).

35. Boumendjel, F., et al. (2016). Study of wild boar density in the Tlemcen region. Revue Algérienne d'Écologie, 45(2), 89-97.
36. Boumendjel, F.Z , Hajji,G , Valqui,J & Bouslama,Z.(2016).The Hunting Trends of Wild Boar (Sus scrofa) Hunters in Northeastern Algeria, Wildl. Biol. Pract, 12(2): 1-14p.doi:10.2461/wbp.2016.12.9
37. Brahmi K, Ouelhadj A, Baziz B, Doumandji S (2014). Diet ecology of common genet in Bouzeguène moutains (Kabyly, Algeria). Lebanese Science Journal, 15(1), 27-39.
38. Brito, J. C., Godinho, R., Martínez-Freiría, F., Pleguezuelos, J. M., Rebelo, H., Santos, X., Vale, C. G., Velo-Antón, G., & Boratyński, Z. (2009). The impact of road expansion and predator control programs on the golden wolf (Canis anthus) in the Sahara Desert. African Journal of Ecology, 47(3), 541-548.
39. Brito, J., S. Durant, N. Pettorelli, J. Newby, S. Canney, W. Algadafi, T. Rabeil, P. Crochet, J. Pleguezuelos, T. Wacher, K. de Smet, D. Gonçalves, M. da Silva, F. Martínez-Freiría, T. Abáigar, J. Campos, P. Comizzoli, S. Fahd, A. Fellous, H. Garba, D. Hamidou, A. Harouna, M. Hatcha, A. Nagy, T. Silva, A. Sow, C. Vale, Z. Boratyński, H. Rebelo, S. Carvalho. 2018. Armed conflicts and wildlife decline: Challenges and recommendations for effective conservation policy in the Sahara-Sahel. Conservation Letters, 11: 1-13.
40. Brunner, H & Coman, B.J. (1974). "The identification of mammalian hair". Inkata press. Melbourne, Australia,176.
41. Burgin, C.J., Colella, J.P., Kahn, P.L. & Upham, N.S. (2018). How many species of mammals are there? J. Mammal. 99: 1-14.
42. Caby, R., & al. (2001). Geodynamic evolution of the Edough Peninsula. Journal of African Earth Sciences, 32(3), 615-624.
43. Cadhla, F, Meera, B, Matthew, F, Simon.H, Williams, A, Matthew, J, Frye.C, Peter, S, Juliette, M, Conte, J , Joel, G, Nishit, B, Bohyun ,L, Xiaoyu, C, Phenix, L. Q, Ian, L. (2014). "Detection of Zoonotic Pathogens and Characterization of Novel Viruses Carried by Commensal Rattus Norvegicus in New York City." MBio 5(5):1-16.
44. Camp, J. (2012). Generalist and specialist predators in Europe: The case of the European otter and badger. Mammalian Ecology Journal, 18(1), 112-123.
45. Carleton, M.D., and Van der Straeten E., 1997. Morphological differentiation among Sub-Saharan and North-African populations of the Lemniscomys barbarus complex [Rodentia: Muridae]. Proceedings of the Biological Society of Washington, 110: 640-680.
46. Caro, T., & Stoner, C. (2003). The impact of large predator loss on mesopredators and prey communities. Journal of Mammalogy, 84(4), 128-136.
47. Carvalho, J., & Gomes, P. (2004). Genet feeding in an urban environment in Portugal. Journal of Urban Ecology, 2, 45-50.
48. Ceballos, G., & Ehrlich, P. R. (2009). Global mammal conservation: The challenges of the 21st century. Journal of Mammalogy, 90(3), 509-518.
49. Chagas, Carolina Romeiro Fernandes, Irys Hany Lima Gonzalez, Samantha Mesquita Favoretto, and Patrícia Locosque Ramos. 2017. "Parasitological Surveillance in a Rat (Rattus Norvegicus) Colony in São Paulo Zoo Animal House." Annals of Parasitology 63(4):291-97.
50. Chaline, J., Baudvin, H., Jammot, P & Saint-Girons, M. C. (1974). Les proies des rapaces. Doin, Paris.

51. Clutton-Brock, T. H., & Harvey, P. H. (1978). Primate ecology and social organization. In J. G. Else & P. C. Lee (Eds.), Primate ecology: Studies of feeding and ranging behavior in lemurs, monkeys, and apes (pp. 135-148). Cambridge University Press (reprinted in Judas, 1986).
52. Corbet, G. B. (1966). The terrestrial mammals of Western Europe. Oxford University Press.
53. Corbet, G. B. (1989). *The identification of prey items in carnivore scat: A review. Mammal Review*, 19(1), 49-58.
54. COUZI L., 2011 - Identifying the small non-flying mammals, Erinaceomorpha, Soricomorpha, Rodentia of Aquitaine. 24 p.
55. Croquet, C. (2005). Carnivores of North Africa: Historical distribution and conservation. Université de Montpellier Press.
56. Croquet, C. (2005). Conservation status and reproductive strategies of small carnivores in North Africa. European Journal of Wildlife Research, 51(3), 172-181.
57. Cuvier, G. (1825). The animal kingdom distributed according to its organization. Deterville Press.
58. Dahiya, Tejpal. 2020. "Ecologically Based Management Techniques of Rodent Pests." (January).
59. Dalerum, F., Cameron, E. Z., Kunkel, K., & Somers, M. J. (2009). Diversity and depletions in continental carnivore guilds: implications for prioritizing global carnivore conservation. Biology letters, 5(1), 35-38.
60. Dalhoum.R et al. (2018). "Liste Preliminaire Des Mammiferes Terrestres De La Region Mastouta-Bishshouk (Beja, Nord-Ouest De La." 1-10.
61. Damange, J. P. (1999). Technique for the analysis and study of predation in carnivores. Application to martes Foina and Mustela erminea. Diploma of study and research. Univ.de Bourgogne. 41 p.
62. Davison, A., & al. (2002). *Non-invasive monitoring of micromammalian populations through predator diet analysis*. *Conservation Biology*, 16(4), 988-997.
63. Day, M. G. (1966). Identification of haire and feather remains in the gut and faeces of stoats and weasels. J. Zool. Lond. 148: 201-217.
64. De Marinis, A. M., & Asprea, A. (2006). "Hair identification key of wild and domestic ungulates from southern Europe". Wildlife Biology, 12(3), 305-320.
65. Debrot, S. Fivaz, G. Mermod, C. and Weber, J.M. (1982). Atlas des poils de Mammifère d'Europe. Inst. Zool. Univ. Neuchâtel. 208 p.
66. Delibes, M. (1974). Ecology of the common genet in Spain. Mammalia, 38(3), 421-436.
67. Delibes, M. (1979). New subspecies of Genetta genetta from Ibiza Island. Journal of Mammalogy, 60(2), 321-324.
68. Delibes, Rodriguez A. and Parreno,F, (1989). Food of the genet (Genette genetta) in northern African. Est, Biol. Donana, J. Zool, 218, 321-328.
69. De-Marinis, A. M, & Paolo, A. (1993). "Guide to the microscope analysis of Italian mammals hairs: Insectivora, Rodentia and Lagomorpha". Italian Journal of Zoology 60.2. 225-232.
70. Denys, C, Stoetzel, E, Lalis, A, Nicolas,V , Delapre, A, Mataame ,A, Tifraouine, L , Rihane,A , EL Brini , Liefrid ,S Fahd,S ,Ouarour,A , Cherkaoui,A ,Fekhaoui,M , Benhoussa,A , EL Hassani, A and Benazzou, T .(2015). Inventaire des petits mammifères de milieux anthropisés et naturels du Maroc septentrional, Travaux de l'Institut Scientifique, Série Générale, 2015, N° 8, 113-126.

71. Denys, C. (2011). Rodents. In Paleontology and geology of Laetoli: Human evolution in context (pp. 15-53). Springer, Dordrecht.
72. Desmet, K. & Hamdine, W. (1988). Genet (Genetta genetta Linné, 1758) densities in Algerian yeuseraie. Mammalia, 52 (4): 604-607.
73. Drouilly, M., & al. (2018). *Monitoring micromammal populations using predator diet analysis in conservation management. Biodiversity and Conservation*, 27(12), 3215-3230.
74. Dufour, S. (2012). Reviving the African wolf Canis lupus lupaster in North and West Africa: a mitochondrial lineage ranging more than 6,000 km wide. PLoS ONE, 7: e42740.
75. Durant, S., M. Craft, R. Hilborn, S. Bashir, J. Hando, L. Thomas. 2011. Long-term trends in carnivore abundance using distance sampling in Serengeti National Park, Tanzania. Journal of Applied Ecology, 48: 1490-1500.
76. Eddine, A., N. Mostefai, K. de Smet, D. Klees, H. Ansorge, Y. Karssene, C. Nowak, P. van der Leer. 2017. Diet composition of a newly recognized Canid species, the African golden wolf (Canis anthus), in Northern Algeria. Annales Zoologici Fennici, 54: 347-356.
77. Éditions Delachaux et Niestlé, Lausanne, 304 p.
78. Editions Porte-Plumes, Ayer, 309 p.
79. Elbroch, M. (2003). *Mammal Tracks & Sign: A Guide to North American Species*. Stackpole Books.
80. Erome, G et Aulagnier, S.(1982).Cotributions à l'identification des preies de rapaces, Bievre ,4 :129-135.
81. Ewer, R. 1973. Carnivores. Ithaca: Cornell University Press.
82. Farhi, Y., K. Hani, M. L. Ahmat, K. E. Bambra, T. Radjah, K. Absi, and K. Souttou. 2016. "Première Données Sur Le Comportement Trophique de La Chouette Effraie (Tyto Alba Scopoli , 1769) Dans La Région de Biskra (Sahara Septentrionale Algérien)." 13:113-20.
83. Fedriani, J. & Travaini, A. (2000). Predator trophic guild assignment: the importance of the method of diet quantification. Rev. Eclo (Terre Vie). 55: 129-139.
84. Firth, Cadhla, Meera Bhat, Matthew A. Firth, Simon H. Williams, Matthew J. Frye, Peter Simmonds, Juliette M. Conte, James Ng, Joel Garcia, Nishit P. Bhuva, Bohyun Lee, Xiaoyu Che, Phenix Lan Quan, and W. Ian Lipkin. 2014. "Detection of Zoonotic Pathogens and Characterization of Novel Viruses Carried by Commensal Rattus Norvegicus in New York City." MBio 5(5):1-16.
85. Fons R. (1984a) - The shrew (Crocidura russula). In: A. Fayard (ed.): Atlas des mammifères sauvages de France. Société Française pour l'Etude et la Protection des Mammifères, Paris, p. 42-43.
86. Fuller, T. K., & Roughton, R. D. (1989). The ecology and management of golden wolves (Canis anthus) in North Africa. In J. L. Gittleman (Ed.), Carnivore behavior, ecology, and evolution (Vol. 1, pp. 179-196). Cornell University Press.
87. Gasperetti, J., Harrison, D. L., Büttiker, W., & Nader, I. A. (1985). Mammals of Saudi Arabia, vol. 7: Carnivores of Arabia. Fauna of Saudi Arabia Publications.
88. Gaubert, P. (2009). Genets: Taxonomy, evolution, and distribution. Springer-Verlag.
89. Gaubert, P., Bloch, C., Benyacoub, S., Abdelhamid, A., Pagani, P., Djagoun, C.A.M.S., Couloux, A., & Genoud M. et Hutterer R. (1990) - Crocidura russula (Hermann, 1780) - Hausspitzmaus.

90. Gaubert, P., Godoy, J. A., Del Cerro, I., & Palomares, F. (2009). Early phases of a successful invasion: mitochondrial phylogeography of the common genet (Genetta genetta) within the Mediterranean Basin. Biological invasions, 11, 523-546.
91. Gomes, P., & Giraudoux, P. (1992). Landscape structure and habitat of the Genet (Genetta genetta, L.). XVth Colloque Francophone de Mammalogie: Les carnivores. 59-63.
92. Gomes, P.T. (1993). Space use by a carnivore Mammal, Genetta genetta, L.: Importance of landscape structure. In: Proceeding IUGB XXI congress, 265-267.
93. Gutema, T. M, Atickem, A, Bekele, A. S. Z, Claudio, K, Mohammed, T, Diress, V, Vivek, V. Fashing, P. J, Zinner, D, Stenseth, N. C. (2018). "Competition between sympatric wolf taxa: An example involving African and Ethiopian wolves". Royal Society Open Science journal, vol 5, issue 5.
94. Hadj Zobir, A., & al. (2007). Geological and geomorphological characteristics of the Edough massif. Journal of Earth Sciences, 12(4), 345-357.
95. Hadj Zobir, M. (2012). Analysis of the hydrographic characteristics of the Edough peninsula: a dendritic drainage under steep slope. Revue Algérienne de Sciences de la Terre, 20(1), 34-48.
96. Hadjoudj, Moussa, Karim Souttou, and Salaheddine Doumandji. 2015. "Diversity and Richness of Rodent Communities in Various Landscapes of Touggourt Area (Southeast Algeria)." Acta Zoologica Bulgarica 67(3):415-20.
97. Haines, A., R. S. Kovats, D. Campbell-Lendrum, and C. Corvalan. 2006. "Climate Change and Human Health: Impacts, Vulnerability and Public Health." Public Health 120(7):585-96.
98. Haltenorth, T., & Diller, H. (1980). Mammals of Africa: A field guide. Collins.
99. Hamdine, W. (1991). Ecology of the genet (Genetta genetta) in the Djurdjura National Park. Station: Tala-Guilef. Magister thesis, I. N. A. El-Harrach : 105 p
100. Hamdine, W., Thevenot, M., Sellami, M. and De Smet, K. (1993). The diet of the Genet (Genetta genetta Linné, 1758) in the Djurdjura National Park, Algeria. Mammalia, 57(1): 9 - 19.
101. Hamel, C., & al. (2022a). Biodiversity analysis in the Edough region, Algeria. Journal of Environmental Studies, 14(1), 45-60.
102. Hamel, C., & al. (2022b). Survey of endemic and sub-endemic taxa in Algeria. Biodiversity Journal, 6(3), 100-115.
103. HANI, A, Djabri, L, Jacky, M. (1997). "Etude des caractéristiques physico-chimiques du massif cristallophyllien de Séraïdi (nord-est Algérien)". IAHS Publications-Series of Proceedings and Reports-Intern Assoc Hydrological Sciences, 241, 47-62.
104. Hani, A., & al. (1995). Carte géologique des grands traits structuraux et colonne lithologique du massif de l'Edough. Mémoires de la Société Géologique de France, 172, 45-59.
105. Hani, A., & al. (1997). Geography and ecology of the Edough massif in Algeria. Revue Algérienne de Géographie, 15(3), 56-72.
106. Harrison, D. L. (1968). The mammals of Arabia. Ernest Benn Limited.
107. Henschel, P, Hunter, L, Breitenmoser, U, Purchase, N, Packer, C, Khorozyan, I, Bauer, H, Marker, L, Sogbohossou, E, Breitenmoser, W. C. (2008). Panthera pardus. In: (IUCN (2011). "Red List of Threatened Species"). [archive]. www.iucnredlist.org

108. Hubert, B, Dominique, G, François, A. (1981). "Annual diet cycle of the three main rodent species (Rodentia; Gerbillidae and Muridae) from Bandia (Senegal). "Mammalia. 1-20.
109. Ikhlef.M & Boughedda.K. (2018). Contribution à l'étude du régime alimentaire et de l'occupation spatiale de la Genette (Genetta genetta L.1758) dans la région d'Ait Zellal (Mekla), Tizi-Ouzou. Master's thesis in animal ecology, univ. Mouloud Mammeri, Tizi Ouzou, Algeria :51PP Annexe.
110. J. Niethammer and F. Krapp (eds.): Handbuch der Säugetiere Europas. Band 3/1, Insektenfresser-Insectivora / Herrentiere-Primates. Aula-Verlag, Wiesbaden, pp. 429-452.
111. Jędrzejewska, B. & Jędrzejewski, W. (1998). Predation in Vertebrate Communities: The Bialowieza Primeval Forest as a Case Study. Ecological Studies 135. 463 p.
112. Judas, J. (1986). Social organization and feeding strategy in carnivores. Mammalia, 50(2), 183-198
113. Judas, J. (1998). Ecology and behavior of the golden wolf in Algeria. Mammalia, 62(2), 215-230.
114. Karim, Souttou, Sekour Makhlouf, Gouissem Kheira, and Hadjoudj Moussa. 2012. "ECOLOGICAL PARAMETERS OF RODENTS SURVEYED IN A SEMI-ARID ENVIRONMENT IN DJELFA (ALGERIA)." 2:28-41.
115. Karssene Y., Chammem M., Li F., Eddine A., Hermann A., Nouira S.(2019). Spatial and temporal variability in the distribution, daily activity and diet of fennec (Vulpes zerda), red fox (Vulpes vulpes) and African golden wolf (Caniss anthus) in southern Tunisia. Mamm. Biol. 95: 41-50.
116. Khidas, K. (1986). Étude de l'organisation sociale et territoriale du chacal Canis aureus Wagner 1841, dans le Parc National du Djurdjura. Université des Sciences et de la technologie d'Alger, Algiers.
117. Khidas, K. (1990). Reproductive biology of the golden wolf (Canis anthus) in North Africa. Journal of Zoology, 23(4), 345-358.
118. Khidas, K. (1998). Seasonal reproduction in golden wolves: Observations from Algeria. African Journal of Zoology, 29(1), 200-207.
119. Khidas, K. 1993. "Distribution Des Rongeurs En Kabylie Du Djurdjura (Algérie)." Mammalia 57(2):207-12.
120. Khidas, K. 1998. Distribution and habitat selection norms in terrestrial mammals of the Djurdjura Kabylia. Doctoral thesis in Biology. Univ. de Tizi-Ouzou, Tizi-Ouzou 235p.
121. Khidas, K., Nora Khammes, Samia Khelloufi, S. Lek, and S. Aulagnier. 2002a. "Abundance of the Wood Mouse Apodemus Sylvaticus and the Algerian Mouse Mus Spretus (Rodentia, Muridae) in Different Habitats of Northern Algeria." Mammalian Biology 67(1):34-41.
122. Khidas, K., Nora Khammes, Samia Khelloufi, S. Lek, and S. Aulagnier. 2002b. "Abundance of the Wood Mouse Apodemus Sylvaticus and the Algerian Mouse Mus Spretus (Rodentia, Muridae) in Different Habitats of Northern Algeria." Mammalian Biology.
123. Kingdon, J. (2010). The Kingdon field guide to African mammals. A&C Black.
124. Klare, U. N. N., Kamler, J. F., Stenkewitz, U. T. E., & Macdonald, D. W. (2010). Diet, prey selection, and predation impact of black-backed jackals in South Africa. The Journal of Wildlife Management, 74(5), 1030-1041.

125. Kowalski, KAZIMIERZ and BARBARA Rzebik-Kowalska. 1991. "Mammals of Algeria." 353.
126. Krebs, C. J., & al. (2014). *Ecology of small mammal populations: Strategic positioning within food webs. Population Ecology*, 56(1), 1-12.
127. Krebs, C. J., Cowcill, K., Boonstra, R., & Kenney, A. J. (2010). Do changes in berry crops drive population fluctuations in small rodents in the southwestern Yukon? Journal of Mammalogy, 91(2), 500-509.
128. LAREF, N., REZZAG-BEDIDA, R. A. N. I. A., BOUKHEROUFA, M., SAKRAOUI, R., HENADA, R. L. I., HADIBY, R., & SAKRAOUI, F. (2022). Diversity and status of day butterflies (Lepidoptera: Rhopalocera) in different plant associations of the Edough Forest Massif (Northeastern Algeria). *Biodiversitas Journal of Biological Diversity*, *23*(2).
129. Le Berre, M. (1990). Carnivores of the Sahara and North Africa. Société des Amis du Muséum National d'Histoire Naturelle.
130. Le Jacques D., Lodé T.(1994). The diet of the European Genet, Genetta genetta L., 1758, in a bocage in western France. Mammalia. 58 (3): 383-389.
131. Léger, F., & Ruette, S. (2010). Reproduction strategies in martens and genets: A comparison. European Mammal Biology, 25(3), 110-118.
132. Linnaeus, C. (1758). Systema Naturae per Regna Tria Naturae. Laurentii Salvii.
133. Livet, A., & Roeder, J. J. (1987). The reproductive biology of the common genet (Genetta genetta). Mammalia, 51(2), 178-184.
134. Livet, V., & Roeder, J. J. (1987). Distribution and habitat of the genet (Genetta genetta) in France. Mammalia, 51(4), 563-570.
135. Lodé, T. (1989). Dynamique des relations trophiques de Mustela putorius et de ses preies : signification adaptative des variabilités interindividuelles des stratégies de prédation. Doctoral thesis, Université de Rennes I. 235 p.
136. Lodé, T., Lechat, I., & Le Jacques, D. (1991). The diet of the genet at the northwestern limit of its range. Rev. Ecol. (Terre Vie). 46 : 339-348.
137. Lozé, I. (1984). Diet and use of space in the genet (Genetta genetta). D. E. A. in behavioral biology, Université de Paris VIII.
138. Lugon-Moulin N. (2003) -Shrews. Biology, ecology, distribution in Switzerland.
139. Macdonald, D. W. (1980). Patterns of scent marking with faeces among carnivore communities. Journal of Zoology, 192(4), 557-564.
140. Macdonald, D. W. (2006). The biology and conservation of wild canids. Oxford University Press.
141. Maizeret, C., & al. (1990). Habitat and behavior of the genet in southern France. Revue de Mammalogie, 3(1), 91-99.
142. Makundi, Rhodes H. and Apia W. Massawe. 2011. "Ecologically Based Rodent Management in Africa: Potential and Challenges." Wildlife Research 38(7):588-95.
143. Mallil,K.(2012). Comparaison des caractéristiques du régime alimentaire et de l'occupation de l'espace de la Genette (Genetta genetta L.1758) (Genetta genetta L.1758) Genetta genetta L.1758) dans deux milieux du Nord Algerien dans deux milieux du Nord algérien algérien: Parcs Nationaux du Djurdjura et d'El: Parcs Nationaux du Djurdjura et d'El: Parcs Nationaux du Djurdjura et d'ElKala. Magisteren Ecologie et Biodiversité Animales des Ecosystèmes Continentaux , Université Mouloud Mammeri de Tizi-Ouzou ,172p

144. Mansour, A., Giraudoux, P., & Delattre, P. (2005). Comparative study of the diet of two sympatric carnivores - the jacal (Canis aureus) and the Genet (Genetta genetta) - at two sites in Kabylia, Algeria. Mammalia. pp. 1 - 32.
145. Mardonald ,D. and Barrett P. (1995) - Guide complet des mammifères de France et d'Europe.
146. Matschie, P. (1902). Die Säugetiere Deutschlands und ihre Stämme. Fischer Press.
147. Mazza, G., & Mori, E. (2023). Solitary, in the night, it goes: summary of records and range expansion of the common genet in Italy. Mammalia, 87(1), 29-33.
148. McFarlane, Ro, Adrian Sleigh, and Tony McMichael. 2012. "Synanthropy of Wild Mammals as a Determinant of Emerging Infectious Diseases in the Asian-Australasian Region." EcoHealth 9(1):24-35.
149. Mech, L.D., & Boitani, L. (2003). *Wolves: Behavior, Ecology, and Conservation*. University of Chicago Press.
150. Meerburg, Bastiaan G., Grant R. Singleton, and Aize Kijlstra. 2009. Rodent-Borne Diseases and Their Risks for Public Health Rodent-Borne Diseases and Their Risks for Public Health. Vol. 35.
151. Mills, L. S., & Knowlton, F. F. (1991). Coyote space use in relation to prey abundance. Canadian Journal of Zoology, 69(6), 1516-1521.
152. Mistrot, V. (2000). Micromammals, markers of environmental anthropization. Études rurales, (153-154), 195-206.
153. Moehlman, P. 1987. Social organization in jackals: The complex social system of jackals allows the successful rearing of very dependent young. American Scientist, 75: 366-375.
154. Moehlman, P. D. & Jhala. Y. V. 2013. Golden jackal (Canis aureus). - In: Kingdon. J. and Hoffman, M. Mammals of Africa, Bloomsbury, London, 5, 35-38.
155. Moehlman, P. D. (1987). Behavior and reproduction in the golden wolf. Animal Behaviour, 35(3), 1121-1130.
156. Moehlman, P. D., & Hayssen, V. (2018). Canis anthus: Life history, ecology, and conservation. Smithsonian Contributions to Zoology, 48(2), 200-215.
157. Moehlman, P., V. Hayssen. 2018. Canis aureus (Carnivore: Canidae). Mammalian Species, 50: 14-25.
158. Monat, S., & Pustoch, V. (2001). *Dental formula and morphology of the shrew (Soricidae): Implications for phylogeny and ecology*. In *Mammalian Biology*, 66(1), 57-69.
159. Naves, J., Wiegand, T., Revilla, E., & Delibes, M. (2003). Endangered species constrained by natural and human factors: the case of brown bears in northern Spain. Conservation biology, 17(5), 1276-1289.
160. Nicolas, Violaine, Arame Ndiaye, Touria Benazzou, Karim Souttou, Arnaud Delapre, and Christiane Denys. 2014. "Phylogeography of the North African Dipodil (Rodentia: Muridae) Based on Cytochrome- b Sequences ." Journal of Mammalogy 95(2):241-53.
161. Obiegala A, Król N, Oltersdorf C, Nader J, Pfefer M. (2017).The enzootic life-cycle of Borrelia burgdorferi (sensu lato) and tick-borne rickettsiae: an epidemio logical study on wild-living small mammals and their ticks from Saxony, Germany. Parasit Vectors. 10:115.
162. Oppliger, A. (2008). *Small mammals as environmental indicators: Their role in ecosystem health. Ecological Research*, 23(5), 943-952.

163. Oppliger, J. (2008) - Micromammals (Chiroptera, Insectivora and Rodentia) as environmental indicators in the Tardiglacial and Holocene: the case of Moulin du Roc (Saint-Chamassy, Dordogne, France). Université de Genève, Faculté des sciences, Section de biologie, 130 p. (unpublished) http ://archive-ouverte.unige.ch/unige :15809
164. Ordeñana MA, Crooks KR, Boydston EE, Fisher RN, Lyren LM, Siudyla S, Haas CD, Harris S, Hathaway SA, Turschak GM, Miles AK, Van Vuren DH.(2010). Effects of urbanization on carnivore species distribution and richness, Journal of Mammalogy, 91(6): 1322-1331, http://www.bioone.org/doi/full/10.1644/09-MAMM-A-312.1
165. Oularbi, S., & Zeghiche, H. (2009). Impact of precipitation on erosion in the Edough peninsula. Bulletin de l'Institut Scientifique, 41(3), 89-97.
166. Palazón, S., & Rafart, E. (2010). Dieta de la gineta común Genetta genetta (Linnaeus, 1758) en los hábitats riparios de Navarra. Galemys, 22(2), 3-18.
167. Palomares, F., Rodríguez, A., Laffitte, R., & Delibes, M. (1991). The status and distribution of the Iberian lynx Felis pardina (Temminck) in Coto Donana area, SW Spain. Biological Conservation, 57(2), 159-169.
168. Pianka, E. R. (1973). *The structure of lizard communities*. In *Annual Review of Ecology and Systematics*, 4, 53-74.
169. Poché, R. M., & Ménard, N. (1987). Habitat preferences of golden wolves in Algeria. Wildlife Journal, 21(3), 275-287.
170. Pringle, J. A. (1977). The carnivores of southern Africa. Purnell Publishing.
171. Purchart, L, Suchomel, J, Ladislav, Šipoš, J, Dokulilová, M, C, Heroldová, M, Cepelka, L. (2020). "Effect of Rodents on Forest Regeneration in Clearcuts." Conference: FORESTS' FUTURE: Consequences of Bark Beetle Calamity for the Future of Forestry in Central Europe. Jihlava
172. Purchart, Lubos and Josef Suchomel. 2020. "Effect of Rodents on Forest Regeneration in Clearcuts." (April):2-3.
173. Quéré, J. P. (1993). *Taxonomic keys for the identification of micromammal remains from predator faeces. Mammalia*, 57(1), 21-31.
174. Rabiee, Mohammad Hasan, Ahmad Mahmoudi, Roohollah Siahsarvie, Boris Kryštufek, and Ehsan Mostafavi. 2018. "Rodent-Borne Diseases and Their Public Health Importance in Iran." PLoS Neglected Tropical Diseases 12(4):1-20.
175. Radford, J., & al. (2011). Plant biodiversity in the Edough massif. Flora Mediterranea, 21, 15-30.
176. Raven, P. H. (1963a). The generic position of "Boisduvalia iasmanica Aliso 5; 247-249. . 1963". Amphitropical relationships in the floras of North and South America. Quart. Rev. Biol. 38: 15.
177. Reinthal, P. N. (1990). *Trophic structure and resource use in lizard communities: A test of Pianka's niche overlap model*. In *Ecology*, 71(3), 881-887.
178. Reshamwala, H. S, Shrotriya, S. Bora, Bhaskar, L, Salvador, D. R, Habib, B. (2018). "Anthropogenic food subsidies change the pattern of red fox diet and occurrence across Trans-Himalayas, India". Journal of Arid Environments, vol 150, issue 5,p 15-20.
179. riagno, J. (1985). The habitat of the genet in France. Cahiers de la Société Française d'Écologie, 10(2), 23-35.
180. Rigaux P. & Dupasquier C. (2012). Identification key "in hand" of the micromammals of metropolitan France. Société Française pour l'Étude et la Protection des Mammifères d'Alsace : 56 p.

181. Ritto, J. C, Durant, N, Pettorelli, J, Newby, S, Canney, W, Algadafi, T, Rabeil, P, Crochet, J, Pleguezuelos, T, Wacher, K, De-Smet, D, Gonçalves, M, Da-Silva, F, Martínez-Freiría, T, Abáigar, J, Campos, P, Comizzoli, S, Fahd, A, Fellous, H, Garba, D, Hamidou, A, Harouna, M, Hatcha, A, Nagy, T, Silva, A, Sow, C, Vale, Z, Boratyński, H, Rebelo, S. Carvalho (2018). "Armed conflicts and wildlife decline: Challenges and recommendations for effective conservation policy in the Sahara-Sahel".Conservation Letters, 11: 1-13.
182. Roberts, P. D., Somers, M. J., White, R. M., & Nel, J. (2007). Diet of the South African large-spotted genet Genetta tigrina (Carnivora, Viverridae) in a coastal dune forest.
183. Roeder, J. J. (1980). Study on the defecation sites of Genetta genetta in rocky environments. Mammalia, 44(3), 343-350.
184. Rolland, M. (2008). *Characterization of small mammal species: study of unicupsids and their variation*. *Journal of Mammalogy*, 89(2), 123-130.
185. Rolland,C. (2008). Clé des micromammifères de Rhône-Alpes Identification à partir des restes osseux contenus dans les pelotes de réjection des rapaces, 54p
186. Rosalino, L., & Santos-Reis, M. (2002). Habitat and diet of the Genet in Portugal. European Journal of Wildlife Research, 48(1), 25-30.
187. Rueness, E. K., Asmyhr, M. G., Sillero-Zubiri, C., Macdonald, D. W., Bekele, A., Atickem, A., & Stenseth, N. C. (2011). The cryptic African wolf: Canis aureus lupaster is not a golden jackal and is not endemic to Egypt. PLoS One, 6(1), e16385.
188. Ruiz-Olmo, J. & Lopez-Martin, J.M. (1993). Note on the diet of the common genet (Genetta genetta L.) in Mediterranean riparian habitats of N.E. Spain. Mammalia, 57: 607-610.
189. Sadler, J. P. (2012). *Reproductive cycles in small mammals: Seasonal patterns and environmental influences*. *Journal of Mammalian Biology*, 77(3), 234-246.
190. Sanchez, M. & Rodrigues, P. (2008). Feeding habits of the genet Genetta genetta in an Iberian continental wetland. Hystrix (It. J. Mamm.), 19: 133-142.
191. Sari, A., & Arpacik, A. (2018). Morphological hair identification key of common mammals in Turkey. *Applied Ecology and Environmental Research*, *16*(4), 4593-4603.
192. Schlawe, L. (1980). A review of the genus Genetta in Africa. African Journal of Zoology, 15(3), 180-198.
193. Schlawe, L. (1981). Ecology of North African carnivores. African Wildlife Press.
194. Schlawe, L. (1981). The taxonomy of African viverrids: Focus on Genetta. Mammalia Africana, 14(2), 205-215.
195. Science, N. (2009). "Special publications". IEEE Spectrum (Vol. 13, Issue 4).
196. Souret L. & Riols C. (2018) Étude du régime alimentaire de la Genette commune (Genetta genetta) et de sa répartition en région Sud - Provence-Alpes-Côte d'Azur. Faune PACA Publication, n° 79 : 53 p.
197. Souret, G., & Riols, F. (2018). Genet feeding in French forests. Journal of Mammalogy, 99(3), 784-790.
198. Spitz, F. (1963). Echantillonnage des populations de petits mammifères. Revue d'Écologie (La Terre et La Vie), 17(2), 203-237.
199. Storch, I., Lindstrom, E. & De Jounge, J. (1990). - Diet and habitat selection of the pine marten in relation to competition with the red fox. Acta Theriol, 35: 311-320
200. Stoyanov, S. (2020). Cranial variability and differentiation among golden jackals (Canis aureus) in Europe, Asia Minor and Africa. ZooKeys, 917, 141.

201. T, Aalbu, R, Adlard, R, Adriaenssens, E. M, Aedo, C, Aescht, E, Akkari, N, Alexander, S (2022). Catalog of Life Checklist (Version 2022-12-19). Catalog of Life. https://doi.org/10.48580/dfqt Genetta G.Cuvier, 1816 Published in: Règne Anim. vol. 1 p. 156
202. Tamnling, T., & al. (2012). *Predator-prey interactions and their implications for micromammal monitoring. Ecological Indicators*, 13(1), 104-110.
203. Tanguy, A., & Gourdain, P. (2011). Guide méthodologique pour les inventaires faunistiques des espèces métropolitaines terrestres (volet 2)-Atlas de la Biodiversité dans les Communes (ABC). Muséum National d'Histoire Naturelle, Ministère de l'Écologie, du Développement durable, des Transports et du Logement.Teshome, Z., & Girmay, T. (2015). changing world". International Journal of Current Research.,Ethiopa.
204. Teshome, Z., & Girmay, T (2015). "Ectoparasites of small mammals: ECTOPARASITES OF SMALL MAMMALS: ecology, infection andmanagement in the changing worl. "International Journal of Current Research. Department of Biology, Adigrat University, P. O. Box, 50, Adigrat, Ethiopia.
205. Thomas, O. (1902). Descriptions of new species of mammals from the Balearic Islands. Proceedings of the Zoological Society of London, 1902, 105-110.
206. Torre I, Arrizabalaga A, Freixas L, Ribas C, Flaquer M, Diaz P (2013). Using scats of a generalist carnivore as a tool to monitor small mammal communities in Mediterranean habitats. Basic and appliedecology,14, 155-164.
207. Torre, I, Arrizabalaga, A , Ribas ,A.(2015).The diet of the genet (Genetta genetta Linnaeus, 1758) as a source of information on local small mammal communities, Galemys, 27: 1-6. DOI: 10.7325/Galemys.2015.N4
208. Torre, I., & al. (2013). Dietary data on the Genet in Spain. Ecología de la Fauna, 18, 30-35.
209. Torres, R. T., & al. (2013). *Small mammal diversity and abundance inferred from predator diets. Journal of Zoology*, 289(1), 42-48.
210. Toubal, A. (2013). Impact of land use on pedogenesis in Edough forests. Journal des Sciences de la Terre, 31(4), 317-328.
211. Toubal, A., & al. (2014). Ecological study of the Guerbès Senhadja wet plain: A Ramsar site in Algeria. Journal of Ecology and Conservation, 32(2), 102-112.
212. Toubal, A., & Boumaza, A. (1989). Characteristics of brown and leached soils of the Edough forests. Revue de l'Institut National Agronomique, 24(2), 45-58.
213. Toubal, O, Boussehaba, A, Toubal, A, Samraoui, B. (2014). "Mediterranean biodiversity and global change: the case of the Guerbès-Senhadja wetland complex (Algeria)". Physio-Geo. Physical geography and environment, Vol 8, 273-295.Purchart,
214. Tristan, A. (1960). The distribution of Genetta genetta in North Africa. In Kowalski, K., & Rzebik-Kowalski, B. (1991), Fauna of Algeria: A parasitological perspective (pp. 145-153). Polish Academy of Sciences Press.
215. Tupinier.Y, 1973 morphologie des poils des chiroptres de l'Europe occidentalepar étude de microscope electroniqueà balyage. Revue suisse zool.80 : 635 - 653.
216. Un site enchanteur en péril [archive], on Djazairess (accessed 09/01/2023). https://www.djazairess.com/fr/liberte/101083 Accessed 09/01/2023.
217. Véla, E., & Benhouhou, S. (2007). Biodiversity hotspots in Algeria: Kabylie - Numidie-Kroumirie. Journal of Biodiversity Conservation, 12(4), 123-135.
218. Vincent, E. (2000). *Micromammals as indicators of ecosystem quality and resource availability. Conservation Ecology*, 14(2), 221-230.

219. Vingada, J., & al. (1993). Diet dynamics of the Genet in Portugal. Mammalian Biology, 58, 319-325.
220. Virgos, E., & al. (1996). Genet diet in rural Spain. Spanish Journal of Ecology, 10, 42-50.
221. Virgos, E., & al. (1999). Seasonal variability of Genet diet in Spain. Journal of Mammalogy, 80(4), 1030-1040.
222. Virgos, E., & Casanova, J. (1997). Habitat preferences of the common genet in Mediterranean ecosystems. Wildlife Biology, 3(2), 127-134.
223. Wacher, T., De Smet, K., Belbachir, F., Belbachir-Bazi, A., Fellous, A., Belghoul, M., & Marker, L. (2005). Faunal inventories of the Sahelo-Saharan interest group. Massif central de l'Ahaggar, Algeria (March 2005). iv+ 40 p.
224. Widdows. Craig, D.Roberts, Peter. D.Maddock, Anthony. H. Carvalho, Filipe. Gaubert, Philippe San, Emmanuel Do Linh, 2016. Genetta tigrina - Cape Genet. The Red List of Mammals of South Africa, Swaziland and Lesotho journal. P1-6.
225. Wilson, D.E, Reeder, D.M.(2005) Mammal Species of the World: A Taxonomic and Geographic Reference. JHU Press. 1426 p.
226. Wilson, E. O., & Reeders, A. (1993). Mammalian predators as indicators of ecosystem health. Biological Conservation, 64(1), 45-58.
227. Winemiller, K. O., & Polis, G. A. (1996). *Food webs: Integration of patterns and dynamics*. Chapman & Hall.
228. Winemiller, K. O., & Polis, G. A. (1996). *Food webs: Integration of patterns and dynamics. Annual Review of Ecology and Systematics*, 27, 311-338.
229. Yahi, A., & al. (2012). Flora and fauna of the Edough massif. Ecological Research in Algeria, 10, 77-88.
230. Yalden, D., M. Largen, D. Kock, J. Hillman. 1996. Catalog of the mammals of Ethiopia and Eritrea. 7. Revised checklist, zoogeography and conservation. Tropical Zoology, 9: 73-164.
231. Zemiti, B. (2012). Do you know the hunting period and phenological status of our game animals? [Do you know the hunting period and phenological status of our game animals? Bulletin d'information et de vulgarisation (La lettre cynégétique). Center Cynégétique de Zéralda. Algiers.

I want morebooks!

Buy your books fast and straightforward online - at one of world's fastest growing online book stores! Environmentally sound due to Print-on-Demand technologies.

Buy your books online at
www.morebooks.shop

Kaufen Sie Ihre Bücher schnell und unkompliziert online – auf einer der am schnellsten wachsenden Buchhandelsplattformen weltweit! Dank Print-On-Demand umwelt- und ressourcenschonend produzi ert.

Bücher schneller online kaufen
www.morebooks.shop

info@omniscriptum.com
www.omniscriptum.com

OMNIScriptum

MIX
Papier aus verantwortungsvollen Quellen
Paper from responsible sources
FSC® C105338

Printed by Books on Demand GmbH, Norderstedt / Germany